T0213554

SpringerBriefs in Materials

The SpringerBriefs Series in Materials presents highly relevant, concise monographs on a wide range of topics covering fundamental advances and new applications in the field. Areas of interest include topical information on innovative, structural and functional materials and composites as well as fundamental principles, physical properties, materials theory and design.

Indexed in Scopus (2022).

SpringerBriefs present succinct summaries of cutting-edge research and practical applications across a wide spectrum of fields. Featuring compact volumes of 50 to 125 pages, the series covers a range of content from professional to academic. Typical topics might include

- A timely report of state-of-the art analytical techniques
- A bridge between new research results, as published in journal articles, and a contextual literature review
- A snapshot of a hot or emerging topic
- An in-depth case study or clinical example
- A presentation of core concepts that students must understand in order to make independent contributions

Briefs are characterized by fast, global electronic dissemination, standard publishing contracts, standardized manuscript preparation and formatting guidelines, and expedited production schedules.

More information about this series at https://link.springer.com/bookseries/10111

Rebeka Rudolf · Vojkan Lazić · Peter Majerič ·
Andrej Ivanič · Gregor Kravanja · Karlo T. Raić

Dental Gold Alloys and Gold Nanoparticles for Biomedical Applications

 Springer

Rebeka Rudolf 🆔
Faculty of Mechanical Engineering
University of Maribor
Maribor, Slovenia

Peter Majerič
Faculty of Mechanical Engineering
University of Maribor
Maribor, Slovenia

Gregor Kravanja
Faculty of Chemistry and Chemical
Engineering
University of Maribor
Maribor, Slovenia

Vojkan Lazić
Faculty of Dentistry
School for Dental Medicine
University of Belgrade
Belgrade, Serbia

Andrej Ivanič
Faculty of Civil Engineering,
Transportation and Architecture
University of Maribor
Maribor, Slovenia

Karlo T. Raić 🆔
Faculty of Technology and Metallurgy
University of Belgrade
Belgrade, Serbia

ISSN 2192-1091   ISSN 2192-1105   (electronic)
SpringerBriefs in Materials
ISBN 978-3-030-98745-9   ISBN 978-3-030-98746-6   (eBook)
https://doi.org/10.1007/978-3-030-98746-6

This Springer imprint is published by the registered company Springer Nature Switzerland AG
The registered company address is: Gewerbestrasse 11, 6330 Cham, Switzerland

Preface

Gold has always been around us, tempting, fascinating and financially desirable.

In time, that wonderful *Aurum* crept into our everyday life, our bodies, as well as neoteric technologies.

Through its mysterious paths, gold has conquered medicine through a handful of areas.

Modern achievements in the application of gold and gold dental alloys are indicated in this occasional monograph, in which the present authors have participated for many years through their intensive scientific and professional work.

Thus, after introductory information on gold, a detailed analysis of Dental Gold Alloys (DGAs) is given in the second chapter. The third chapter points out the importance of gold nanoparticles (GNPs), which are increasingly present in all areas of science and applied technology, with a special emphasis on biomedical applications.

The authors tried to bring this area closer to future readers in an acceptable way and encourage them to further inventive reflections on the impact of gold on our future.

In preparing the monograph, we thank the company Zlatarna Celje d.o.o. from Slovenia, which provided the samples of Au dental alloys and gold nanoparticles. Based on this, it was possible to conduct the research and prepare the appropriate text of the monograph, which will be, we hope, likely useful for readers of fields such as dentistry, materials, advanced technologies, biotechnology and characterisation.

Maribor, Slovenia   Rebeka Rudolf
Belgrade, Serbia   Vojkan Lazić
Maribor, Slovenia   Peter Majerič
Maribor, Slovenia   Andrej Ivanič
Maribor, Slovenia   Gregor Kravanja
Belgrade, Serbia   Karlo T. Raić

Contents

About the Authors

Assoc. Prof. Dr. Rebeka Rudolf after graduation from the Faculty of Mechanical Engineering in the University of Maribor, Slovenia, continued with postgraduate studies (Master's Degree in 1997) up to a Doctorate in 2002. She has been employed permanently at the University since 1993. In 2006, she was additionally employed by Zlatarna Celje (ZC) as Research Manager and as Leader of the Research Group. In 2017, she achieved the level of Research Counsellor and Associate Professor in the University of Maribor (Habilitation field: Materials). She has focused intensively on the development of new gold dental alloys, studying their biocompatibility and corrosion resistance, which are the applicable values for the field of dentistry. Under her leadership, ZC has put four new dental alloys on the EU market, for which they conducted preclinical investigations complying with the Standards EN ISO 10993-1: 2009 and 10993-5: 2009. The dental alloys are also protected by patents. The named Researcher was promoted on 14.9.2010 by the website Nano Patents and Innovations from the USA—dedicated to new materials, patents, markets' products and research innovation, where they demonstrated that the modern technological materials used in a wide range of medical devices and implants are entirely biocompatible and should not cause health problems as has been suggested previously. In the last 5 years, she has been involved actively in the field of nanotechnology—the synthesis of the gold and other different metal nanoparticles based on the ultrasonic spray pyrolysis process—in ZC, the pilot device is currently under construction. Also, in the last 5 years, she has published in co-authorship many top references in the fields of dental alloys and nanotechnology. e-mail: rebeka.rudolf@um.si

Prof. Dr. Vojkan Lazić is Full Professor and Vice Dean at the School of Dental Medicine at the University of Belgrade. After graduation from the School of Dental Medicine in Belgrade in 1991, he continued with a Master of Science Degree in 1998 to a Doctorate in 2003. He has been employed at the University permanently since 1994. Since then, he has been involved in teaching and research, especially in dental materials. The main research is in the field of new dental alloys for ceramic fused to metal restorations, acrylic resin with nanoparticles for complete or partial dentures and silicones for epitheses. He was Participant and Main Researcher in six scientific projects, mainly in dental materials. In July 2008, he was visting Professor at the Advanced Prosthodontic Division at UCLA, CA—USA. Until now, he has published 52 papers, of which 45 were scientific papers, then 110 reports at international and domestic conferences and six books, of which he co-authored in three textbooks. So far, he has been Mentor in six completed Doctoral Dissertations.

Dr. Peter Majerič completed his university education in 2012 at the Faculty of Mechanical Engineering, University of Maribor. He then continued with Doctoral studies at the Institute of Materials Technology at the same faculty. He worked on a production process for the production of gold nanoparticles in collaboration with IME, the Institut für Metallurgische Prozesstechnik und Metallrecycling, RWTH Aachen University in Germany. In 2016, he finished his Ph.D. Dissertation entitled "Synthesis of gold nanoparticles with a modified Ultrasonic Spray Pyrolysis". In the same year, he continued his employment at the Institute of Materials Technology as Assistant with a PhD. He works primarily on research and development projects and programmes in the field of materials, nanotechnology and nanomaterials' production. He also operates and maintains instruments for scanning electron microscopy (SEM) and has experience in metallography and testing of mechanical properties from cooperation on various projects focused on the functional properties of materials. e-mail: peter.majeric@um.si

Assoc. Prof. Dr. Andrej Ivanič is Associate Professor and Vice Dean for Education at the Faculty of Civil Engineering, Transportation Engineering and Architecture at the University of Maribor, where he is included in teaching and research work. He is also involved additionally in the Research Programme P2-0120 at the Faculty of Mechanical Engineering of the University of Maribor. The main interests of his research include composite materials, biomaterials, nanomaterials, metal alloys, concrete, recycled materials and mechanical testing of various materials and structures. He is known for his active collaboration with industry and civil society, with the aim of contributing efficiently to the improvement of the quality of structures and materials and sustainable development. His bibliography consists of 328 units, including 19 scientific papers, 1 invited talk and 42 papers in conference proceedings. e-mail: andrej.ivanic@um.si

Dr. Gregor Kravanja is Researcher with a Doctorate at the Faculty of Civil Engineering, Transportation and Architecture at the University of Maribor, where he is included in teaching. He is also involved additionally in the Research Programme P2-0046 (Separation Processes and Production Design) at the Faculty of Chemistry and Chemical Engineering at the University of Maribor. The main interests of his research include biomaterials' fabrication using supercritical fluids, scaffolds' manufacturing, enzyme kinetics, biomaterials and mechanical testing. He is known for his active collaboration with industry and civil society, with the aim of contributing efficiently to the improvement of the quality of life and sustainable development. His bibliography consists of more than 82 units, including 19 scientific papers, 2 invited talks and 22 papers in conference proceedings. He is also Theme Editor in the International Journal of Functional Biomaterials (ISSN 2079-4983; CODEN: JFBOAD). e-mail: gregor.kravanja@um.si

Prof. Dr. Karlo T. Raić is Professor at the Department of Metallurgical Engineering (DME), Faculty of Technology and Metallurgy, University of Belgrade, Serbia (FTM-UB-Ser). He teaches the following courses: Transport Phenomena in Materials Engineering, Surface Engineering, Metallic Materials in Medicine and Iron and Steel-selected Topics. His positions are as follows: (1996–7) President of the Metallurgical Division at FTM-UB-Ser, (2002–9, 2017–2019) Head of the DME, (2009–2018) Member of the Faculty Council, (2009–2020) Editor in Chief of FTM-UB-Ser editions, (1995–) Board and (2018–) Vice President of AMES (Association of Metallurgical Engineers of Serbia), (2005–) Editor in Chief of AMES editions, (2003–2008), Editor in Chief, Metalurgija-Journal of Metallurgy, (2018–) Editor, Journal Metallurgical and Materials Engineering. His scholarship is as follows: DAAD Braunschweig (2001); Max-Plank-Institute for Metals Research, Stuttgart (2002); Washington State University, Pullman, USA (2003); OeAD Leoben (2009); Erasmus+, TU Košice, Slovakia (2018); Erasmus+, TU Wien, Austria (2019).

His field of expertise is Transport Phenomena in Materials Engineering, with special attention on surface modification and characterisation of metallic materials usable in biomedical applications. His scientific bibliography consists of more than 170 units (peer-review journals, national journals, international and national conference books) and more than eight textbooks. He participates in and coordinates numerous national, as well as international, projects. In addition, he acts as Thesis Adviser and Member of organising committees. He is Reviewer for international and national journals, monographs in the field of new materials and technologies, papers for international and national conferences, as well as grant applications. He has also served as President and/or Member of international and national conference committees. e-mail: karlo@tmf.bg.ac.rs

Chapter 1
Introduction

Gold has found great application in Medicine and Dentistry [1, 2], which evolved in two directions. First, as a built-in element in the human body, which has been made possible by the development of numerous gold alloys. Over time, due to their favourable properties, gold alloys have found wide application as diverse embedding elements in Dentistry [3], (see Chap. 2, Fig. 1.1). Secondly, the flourishing of Nanotechnology created the conditions for the mass production of gold nanoparticles of various qualities with versatile applications in Medicine and Dentistry, as well as cosmetics and food [4–6]. The appearances of various gold nanoparticles are shown in Chap. 3.

1.1 Basic Properties of Gold

Pure gold, with the symbol Au (from the Latin: *Aurum*) and atomic number 79 in the elements group 11, has particular characteristics: High malleability, considerable density (19.32 g/cm^3), excellent corrosion resistance and a shiny metallic yellow colour. Its special electronic configuration controls the optical properties, chemical reactivity and crystal structure, and explains the interest regarding the coexistence of these characteristics. Gold has attributes, like a large number of isotopes or its opacity to x-rays, which ensures that it interacts strongly with electrons in both Scanning and Transmission Electron Microscopy [8, 9]. Both of these characteristics allowed the use of gold in different technological applications, as well as in Medicine and Dentistry.

Electronic structure. The huge corrosion resistance of gold is a result of its high first ionisation potential, 9.2 eV, e.g. higher than silver (7.6) and/or copper (7.7). Gold has the electronic configuration [Xe] $4f^{14}5d^{10}6s^1$. The 4f electrons shield the other orbital electrons from the nuclear charge. This effect is called lanthanide contraction, and causes the atomic radius of the lanthanides to decrease across the period as the

© The Author(s), under exclusive license to Springer Nature Switzerland AG 2022
R. Rudolf et al., *Dental Gold Alloys and Gold Nanoparticles for Biomedical Applications*,
SpringerBriefs in Materials,
https://doi.org/10.1007/978-3-030-98746-6_1

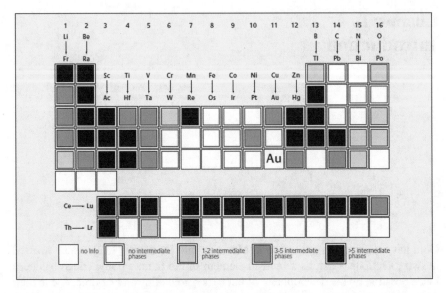

Fig. 1.1 Indication of the number of intermediate phases formed in the binary Au-X systems, for different elements of the periodic table [7]

quality of the shielding per electron decreases. In the 5d metal series this effect gives similar lattice constants to the 4d metals. Also, some relativistic effects are a consequence of this lanthanide contraction effect [8, 10].

Another interesting effect is the "aurophilicity" of gold. It provides an explanation of the interaction between the electrons of nonvalence orbitals, which generates unusual system bonding due to the small energy gap of the 5d and 6 s electrons. Therefore, the normal $5d^{10}$ configuration is easily broken by the hybridisation of the s-d orbitals [11].

Crystal structure and alloying behaviour (compound formation capability). Gold is characterised by a face-centred cubic crystal (fcc) structure, similarly to other ductile metallic elements (e.g. Ag and Cu, which are the most common alloying elements of gold). Contrary to other fcc metals, probably due to its relativistic effects, gold is capable of reconstructing all the three high symmetry planes easily, included in (111). The surface tension of liquid Au is approximately 1.1 to 1.2 J/m², increased by alloying with Cu and, vice versa, lowered with Ag or Zn. The crystal structure of gold, with a multiplicity of close-packed planes on which slip can occur, makes it more ductile in comparison to Fe or W with their bcc structure, or Mg and Zn with their hexagonal structure [12]. However, in gold nanoparticles below 10 nm, the fcc structure is no longer stable, but decahedral, icosahedral, or defects are formed in their structures [13].

Some characteristics, like the electronegativity and the difference in atomic radius, limit the range of solid solutions that gold forms with other elements. Au-Ni, Au-Cu,

Au-Pt and Au–Pd systems form solid solutions at high temperatures, but their mutual solubility decreases as the temperature drops [14].

With other elements gold forms a wide range of intermetallic compounds [7]. The mechanisms that overcome and lead to the formation of intermediate phases in the systems are various: Crystallisation phenomena from the melt, or by ordering in the solid state. Mostly, two sets of elements can be observed, giving intermediate phases. The first set is on the left of the Periodic Table, and corresponds to the formation of stoichiometric compounds with high melting and a strongly negative standard enthalpy of formation (ΔH). The second set of alloys, moving towards the right side of the Periodic Table (e.g. Cd, Ga, In, etc.) generally show low melting points, large solid solubility fields and lower ΔH of formation. A number of these phases are formed through solid-state reactions. Finally, compounds with semi-metals and non-metals are formed at the far right of the Periodic Table, Fig. 1.1 [7].

However, in the greater part of its applications, gold cannot be used pure but, usually, it is necessary to use other metallic alloying elements to ensure suitable properties. The gold alloys' compositions vary over a wide range, due primarily to the need of a particular range of mechanical properties, and often, colours. The additions of particular elements into the gold alloy is very important, for example, in dental applications, where these are used to create a superficial bond with ceramic components and ensuring biocompatibility. Another interesting aspect could be the use of lower melting elements to induce a decrease of melting temperatures, a higher wettability and fluidity of particular gold solders. The various phase diagrams are the basis for any type of metallurgical approach, and both binary and ternary gold alloy phase diagrams have been investigated and evaluated [15–17]. It is possible to correlate some specific aspects of the phase diagrams (the formation of solid solutions, intermediate compounds, etc.) to a number of elemental atomic parameters, generally showing a gradual variation along the Periodic Table. These parameters could be the valence electron number (Periodic Table group number), electronegativity, atomic (and ionic) radius, Mendeleev number, or elemental data such as melting and boiling point, sublimation heat, atomic volume, etc. For instance, reliable data for the solid solubilities of the transition elements in gold indicate that the elements of the 10th and 11th groups (Ag, Cu and Pt, Pd, Ni) present continuous solid solutions, usually at high temperatures [12].

Physical properties. The various physical parameters of gold are reported in Table 1.1 [9]. The melting point of pure Au of 1064.18 °C is one of the calibration points used for the International Temperature Scale ITS-90. The increase of the surface/volume ratio, in small gold particles, generates a decrease in the value of the melting point. This trend changes when the particles become even smaller, about tens of atoms, and the melting point increases again, probably because the elementary thermodynamic model does not consider the molecular regime [18]. Pure gold is characterised by a high thermal and electrical conductivity, albeit lower than silver and copper. The extremely high resistance of this pure metal to corrosion has ensured its place in the connectors required for all types of performance-critical electronic devices. Recent studies have also demonstrated that if the bulk gold is diamagnetic and the magnetic fields repel it weakly, gold nanoparticles (about 2 nm of diameter)

Table 1.1 Various selected physical properties of pure gold, adapted from [9]

Property		Notes
Melting point	1064.18 °C	
Lattice constant	4.07 Å	
Atomic radius	1.44 Å	Metallic radius
Young's modulus	79 GPa	Annealed material at 20 °C
Specific heat	0.1288 J/(g·K) 25.4 J/(Mol·K)	
Electronegativity	2.44	
Electrical resistivity	$2.05 \times 10^{-5} \cdot \Omega \cdot cm$	0 °C
Thermal conductivity	314.4 W/(m·K)	0 °C

are characterised by a not negligible ferromagnetic behaviour. The cause of that could be the particular electronic configuration of their surfaces [19].

Due to the high resistance to oxidation and high surface energy, pure gold can create malleable specimens directly from the molten state. The hardness of pure gold is only 25 HV in the annealed state, which can be raised to about 60–80 HV by application of various cold working processes. Flow stress can rise from about 30–220 MPa by application of 60–70 % cold work. It is unexpected, but well known in surface decoration with gold, e.g. 30 g of gold can be plastic deformed to a sheet of about 25 m^2 in area.

The stacking fault energy of gold has an intermediate value between that of copper and silver. The metals with low stacking fault energies are never looked upon as being particularly ductile. The networks of stacking fault in the gold foil should, theoretically, result in a decrease of ductility, due to the greater difficulty in movement of dislocations. The reason for the high malleability of gold in comparison with other metals characterised by the same fcc structure, is linked to its nobility. Gold, being a noble metal, does not have an oxide film on its surface. Thus, dislocations formed within the gold will be able to escape easily from the metal at the surface. With metals having an oxide film on their surfaces the dislocations could well be held within the metal, and this effect would become more noticeable as the total foil thickness decreased; thus, the flow stress would be increased. In these conditions further strain may be accommodated by sub-grain boundary shearing, so giving rise to fragmentation at a foil thickness equal to the sub-grain diameter. Gold is, however, unusual, in that it can accommodate plastic strain when the beaten foil thickness is less than the sub-grain size produced by plastic straining [20–22].

Optical properties. During metal and light interaction, electrons from the metal surface situated either below or on the Fermi level absorb photons and enter an excited state on or above the Fermi surface respectively. The efficiency of the absorption and re-emission of light depends on the atomic orbitals from which the energy band

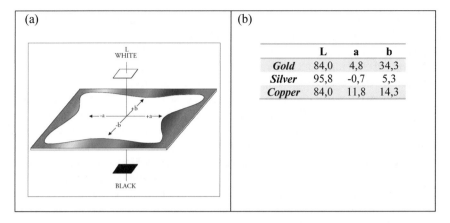

Fig. 1.2 **a** Element of CIELAB **b** examples of L, **a** and **b** values for pure Au, Ag and Cu [24, 25]

originated. Therefore, the procedure for determining the quality of colour is very demanding and complicated [23, 24].

However, the evaluation of the colour, especially gold, is often reduced to individual observation. That was the reason to adopt a new way of measuring and evaluating colours. Today, CIELAB [25] is the most common system, developed by CIE (International Commission on Illumination), and adopted since 1976. It allows describing the colour mathematically without relying on the human eye. The CIELAB method expresses the colour as three-dimensional co-ordinates: L*, a*, and b*, where L* is the luminance (brightness). An L* value of 0 means that no light is reflected by the sample, and an L* value of 100 means that all incident light is reflected. The a* coordinate measures the intensity of the green (negative) or red (positive) component of the spectrum, while the b* coordinate measures the intensity of the blue (negative) or yellow (positive) component. The colour of a sample can be defined by plotting these coordinates as a point in the three-dimensional space represented in Fig. 1.2a. The values of L*, a* and b* of a sample are obtained as direct readings from a spectrophotometer which is connected to a computer. The spectrophotometer has a resolving power between five and ten times greater than the human eye. Finally, an approximate representation of the colour range of a typical base Au alloy (*Au–Ag-Cu*) is given in Fig. 1.2b.

1.2 Aspects of Gold Nanoparticles

Gold nanoparticles (GNPs) have received considerable attention during the past decade due to their potential applications in Catalysis, Chemical sensing, Electronics, Optics, Biology, and especially in Medicine and Dentistry. GNPs are particularly interesting for the production of bio-nanodevices. Their unique optical properties, facile surface chemistry and appropriate size scale are generating much enthusiasm in Molecular biology and Medicine. Their minute size, similar to cellular components and macromolecules, may facilitate the use of GNPs in Medicine for the detection of biological structures and systems, and for the manipulation of cellular functions, as well as in imaging, bio-sensing, and gene and drug delivery.

GNPs have unique chemical and physical properties for transporting and unloading the pharmaceuticals because of their inert and non-toxic core and ease of synthesis. Monodisperse nanoparticles can be formed with core sizes ranging from 1 to 100 nm. Additionally, their photophysical properties could trigger drug release at remote places. Also, monolayer-protected GNPs have recently emerged as an attractive candidate for delivering various therapeutic agents, such as drugs, peptides, proteins and nucleic acids into their targets. GNPs have immense potential for cancer diagnosis and therapy on account of their Surface Plasmon Resonance (SPR), enhanced light scattering and absorption [26, 27].

GNPs for various dental applications

GNPs can serve as a novel application in Dental caries, bone regeneration, Periodontology, Implantology, Tissue engineering and diagnosis of cancer [28–34]. Due to their antibacterial and antifungal actions, they can be used as an additive agent in various dental materials [32].

Dental caries. During the consumption of a carbohydrate rich diet or salivary dysfunction, the action of acidogenic organisms causes breakdown of the enamel that enables the occurrence of caries. GNPs can be used as one of the potential anticaries agents, because of their large nanosize surface area, that allows for more interaction with inorganic and organic molecules.

Dental implants. GNPs act as an osteogenic agent for bone regeneration in two directions: (1) Addition of GNPs enhances the healing of bony defects around an implant, and (2) GNPs enable the conversion of Preosteoblast to Osteoblasts.

Periodontology. To prevent further progression of Periodontal disease and to provide appropriate therapy, diagnosis of Periodontal disease is vital. The unique essential optical features of GNPs make them a key role player in the early and rapid diagnosis of periodontal disease.

Stem cell technology. Nanomaterials are important for Tissue engineering due to their resemblance to a nanostructured type of environment. Thus, nanomaterials can enter the nuclei within the cell, and can affect the functions. In that sense, GNPs have been studied for their effects on stems cells, e.g. Lutz et al. [29] evaluated the application of GNPs on cell matrix adhesions and their effect on the behaviour of cell morphology.

Dental adhesives and composite resins. GNPs did not show any toxic effects on the inhibition of MMP (Matrix metalloproteases) and their cytotoxic responses. So, they are suitable for use as an additional ingredient.

Acrylic resins. The mechanical properties, like the flexural strength and impact fractures strength, were compared between polymer composite PMMA- GNPs (polymethyl methacrylate (PMMA with GNPs) and conventional denture base resins. Incorporation of GNPs improves the mechanical properties of pure PMMA resin (enhances the flexural strength and the thermal conductivity).

Diagnostic imaging in the detection of cancer. GNPs are an effective radio sensitising agent due to a solid photoelectric absorption coefficient in the kilovoltage photon energy range. Thus, they can be used for the detection of cancer at molecular levels. GNPs, either single or conjugated (peptides or antibodies), find application for early detection of cancer [34].

GNPs for use in Lateral Flow Immuno-Assay (LFIA) Tests

Methods, also known as Point Of Care Testing (POCT), are today one of the most commonly used rapid methods of immunological diagnosis, which easily shorten the time between sample collection and analysis. At the same time, they enable the analysis of various samples (blood, saliva, mucus, faeces, urine, etc.) and the measurement of a large number of analytes. POCT devices are advantageous for on-site testing, mainly in terms of (i) Ease of use, (ii) Use of portable instruments, (iii) Elimination of transport of samples to the clinical central laboratory, (iv) Specificity and accuracy, (v) Short sample analysis time, without its pre-treatment, (vi) The use of pre-prepared reagents, and (vii) Reliable results, thus contributing to the rapid identification and management of chronic diseases and acute infections [35–39].

Among POCT devices, devices based on lateral flow immunoassay (LFIA) are used in clinical diagnostics to detect many viruses, diseases and disease states at the site of care. LFIA tests are simple tests, useful for both qualitative and quantitative rapid detection of specific analytes (e.g., antigens and antibodies) in various biological samples. The main reason for their popularity is related to their ease of use, portability, high analytical sensitivity and specificity, and fast results (10–15 min) that can easily be read visually.

The development of highly sensitive and specific LFIA tests in clinical diagnostics has been stimulated strongly in recent years by Nanotechnology, which exploits the exceptional physicochemical properties of metal particles on which depend the visual response of the LFIA test and the associated sensitivity of analysis. Namely, metal particles, in particular, GNPs, which are used as labels for LFIA tests, increase the sensitivity of the LFIA test greatly, due to their active optical and electrical properties and diverse morphology (spheres, cubes, rods, prisms, hollow structures). GNPs are characterised by a special phenomenon called Surface Plasmon Resonance (SPR), which results from the oscillation of free electrons on their surface, which allows GNPs to generate a test signal visible to the naked eye. In addition, GNPs are stable in a variety of media, are biocompatible with a variety of physiological fluids, and have a high surface-to-volume ratio, ensuring their good surface chemistry, i.e., bioconjugation with different molecules. GNPs bind easily to proteins (antibodies, antigens) and, thus, form stable conjugates.

Table 1.2 Advantages and disadvantages of GNPs as labels [37]

Advantages	Disadvantages
High antibody binding affinity (passive binding)	Optimisation of non-antibody protein binding
Possibility of functionalisation of different biomolecules	Possibility of aggregation of larger particles
Vivid and strong colouration	Instability of larger particles
High chemical stability	Poor signal test line due to inadequate GNPs/sample volume ratio
Large specific surface area	Visual multiplexing
Different sizes available (optimal 40 nm)	Slow flow for larger nanoparticles
Easy and cheap synthesis (low cost)	

In the development and production of LFIA tests, GNPs are mostly used as labels, the advantages and disadvantages [37] of which are presented and summarised in Table 1.2.

GNPs-labelled LFIA tests today represent a well-established and extremely versatile diagnostic technology that is used not only in clinical settings, but also in general practice and for home testing. Due to their many properties, which give them the ability to label, the use of LFIA technology has, thus, spread to very different areas of applications, including veterinary medicine, agriculture, the food industry, environmental protection, etc.

LFIA tests can detect both small and large molecules. They can be divided into a number of variants, including their (i) Format, (ii) Recognition molecules (antibodies, aptamers), (iii) Used labels, (iv) Detection systems, and (v) Applications in which they are used. Based on their mode of operation, they can be divided into two main approaches, which have been the most popular in recent decades:

(a) *Direct (sandwich) assay* The direct assay is used to detect and identify relatively large analytes (i.e. high molecular weight analytes), or analytes with multiple binding sites (so-called epitopes). The analyte is trapped between two antibodies (formation of a "sandwich"; see Fig. 1.3. a: One antibody is bound/conjugated to the GNPs, and the other antibody is immobilised on the test line).

The presence of the analyte in the sample will result in colouring of the test line and the test result will be positive. However, when no analyte is present in the sample, no colouring of the test line is observed, and the test result is negative. Any excess of labelled antibodies is captured on a control line, whose colouring is independent of the presence of the analyte in the sample, i.e., the control line is coloured regardless of whether the analyte is present in the sample or not, and ensures that the test is working properly.

For the direct test, the colour or intensity of the signal on the test line is directly proportional to the amount of analyte in the sample (the more analyte, the higher the signal intensity).

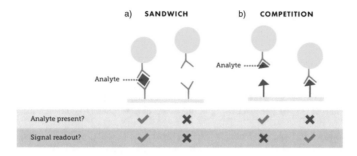

Fig. 1.3 The main configuration of the LFIA test: **a** Direct (sandwich) and **b** Competitive assay

(b) ***Competitive assay*** The competitive assay is used to detect and identify small analytes (i.e., low molecular weight analytes) with a single binding site. In this type of assay, the analyte blocks the binding sites of antibodies that are immobilised on the test line, thus preventing their interaction with the conjugate (see Fig. 1.3b).

If an analyte is present in the sample it binds to the detection antibody, resulting in blockage of the binding site, making such conjugate unable to bind to the test line. The result is an uncoloured test line or lack of signal—the test result is positive. However, if there is no analyte in the sample, there is also no blockage of antibody binding sites—the colour of the test line appears, and the test result is negative. The control line must be visible regardless of the test result.

For a competitive assay the absence of colour on the test line indicates the presence of an analyte, while the appearance of colour on the test line indicates a negative result. The response of the competitive assay is, thus, related negatively to the analyte concentration (the more analyte, the lower the signal intensity, and vice versa).

Each of the approaches has its advantages and disadvantages that we must consider when choosing the right approach of analysis. The advantages and disadvantages depend mainly on the analyte sought, the antibodies used, the sample matrix and the concentration range of interest.

References

1. Corti C, Holliday R (2009) In: Gold: science and application, Taylor & Francis Group
2. Gold Survey (2008) Published by GFMS Ltd. London
3. Knosp H, Holliday RJ, Corti CW (2003) Gold in dentistry: alloys, uses and performance. Gold Bull 36(3):93–102
4. Mody VV, Siwale R, Singh A, Mody HR (2010) Introduction to metallic nanoparticles. J Pharm Bioallied Sci 2:282–289. https://doi.org/10.4103/0975-7406.72127
5. Giljohann DA, Seferos DS, Daniel WL, Massich MD, Patel PC, Mirkin CA (2010) Gold nanoparticles for biology and medicine. Angew Chemie Int Ed 49:3280–3294. https://doi.org/10.1002/anie.200904359

6. Bansal SA, Kumar V, Karimi J, Singh AP, Kumar S (2020) Role of gold nanoparticles in advanced biomedical applications. Nanoscale Adv 2:3764–3787. https://doi.org/10.1039/d0na00472c

7. Ferro R, Saccone A, Maccio D, Delfino S (2003) A survey of gold intermetallic chemistry. Gold Bull. 36(2):39

8. Lide DR (ed) (2008) CRC handbook of chemistry and physics. Taylor & Francis, London

9. Cohn JG (1979) Selected properties of gold. Gold Bulletin 12:21

10. Schmidbaur H (1990) The fascinating implications of new results in gold chemistry. Gold Bull 23:11

11. Schwerdtfeger P, Dolg M, Schwarz WHE, Bowmaker GA, Boyd PDW (1989) Relativistic effects in gold chemistry. I. Diatomic gold compounds. J Chem Phys 91:1762

12. Pyykko P, Desclaux J-P (1979) Relativity and the periodic system of elements. Acc Chem Res 12:276

13. Koga K, Ikeshoji T, Sugawara K (2004) Size and temperature-dependent structural transitions in gold nanoparticles. Phys Rev Lett 92:115507

14. Jansen M (2008) The chemistry of gold as an anion. Chem Soc Rev 37:1826

15. Okamoto H, Massalski TB (eds.) (1987) Phase diagrams of binary gold alloys. ASM International, Metals Park, Ohio

16. Effenberg G, Ilyenko S (eds) (2006) Ternary alloy systems, subvolume B: noble metal systems, landolt-bornstein new series group IV, vol 11. Springer, Berlin, Germany

17. Villars P, Mathis K, Hulliger F, De Boer FR, Pettifor DG (eds.) (1989) In: The structures of binary compounds. Cohesion and structure, vol 2. North Holland, Amsterdam

18. Soule de Bas B, Ford MJ, Cortie MB (2006) Melting in small gold clusters: a density functional molecular dynamics study. J Phys: Condensed Matter 18:55

19. Yamamoto Y, Miura T, Suzuki M, Kawamura N, Miyagawa H, Nakamura T, Kobayashi K, Teranishi T, Hori H (2004) Direct observation of ferromagnetic spin polarization in gold nanoparticles. Phys Rev Lett 93:116801

20. Ott D, Raub ChJ (1981) Grain size of gold and gold alloys. Gold Bull 14(2):69

21. Humphreys FL, M Hatherly (2004) The mobility and migration of boundaries. In Recrystallization and related annealing phenomena. 2nd edn. Elsevier, New York, pp 121

22. Ott D, Raub ChJ (1982) Influence of small additions on the properties of gold and gold alloys (in German), Part I: Metall, 1980, 34, 629; Part II: Metall, 1981, 35, 543; Part III: Metall, 1981, 35, 1005; Part IV: Metall. 36:150

23. German RM, M. Guzowski M, Wright DC (1980) The colour of Au-Ag-Cu alloys: quantitative mapping of the ternary diagram. Gold Bull 13:113

24. Cretu c, van der Lingen E (1999) Coloured gold alloys. Gold Bull 32(4):115–126

25. CIE Colorimetry 15 (2004) 3rd edn. CIE. ISBN 3–901–906–33–9

26. Adekoya JA, Ogunniran KO, Siyanbola TO, Dare EO, Revaprasadu N (2018) Band structure, morphology, functionality, and size- dependent properties of metal nanoparticles. In: Seehra MS, Bristow AD (eds) Noble and precious metals—properties, nanoscale effects and applications. IntechOpen, pp 15–42

27. Bapat RA, Chaubala TV, Dharmadhikarib S, Abdullac AM, Bapat P, Alexandere A, Dubeyf SK, Kesharwanig P (2020) Recent advances of gold nanoparticles as biomaterial in dentistry. Int J Pharmaceut 586:119596

28. Bapat RA, Joshi CP, Bapat P, Chaubal TV, Pandurangappa R, Jnanendrappa N, Gorain B, Khurana S, Kesharwani P (2019) The use of nanoparticles as biomaterials in dentistry. Drug Discov Today. https://doi.org/10.1016/j.drudis.2018.08.012

29. Lutz R, Pataky K, Gadhari N, Marelli M, Brugger J, Chiquet M (2011) Nano-stenciled RGD-gold patterns that inhibit focal contact maturation induce lamellipodia formation in fibroblasts. PLoS One 6:e25459. https://doi.org/10.1371/journal.pone.0025459

30. Yu Q, Li J, Zhang Y, Wang Y, Liu L, Li M (2016) Inhibition of gold nanoparticles(AuNPs) on pathogenic biofilm formation and invasion to host cells. Sci Rep 6:26667

31. Marsh PD (2006) Dental plaque as a biofilm and a microbial community—implications for health and disease. BMC Oral Health BioMed Central. pp S14. https://doi.org/10.1186/1472-6831-6-S1-S14

32. Xia Y, Chen H, Zhang F, Bao C, Weir MD, Reynolds MA, Ma J, Gu N, Xu HHK (2018) Gold nanoparticles in injectable calcium phosphate cement enhance osteogenic differentiation of human dental pulp stem cells. Nanomedicine 14:35

33. Hashimoto M, Sasaki JI, Yamaguchi S, Kawai K, Kawakami H, Iwasaki Y, Imazato S (2015) Gold nanoparticles inhibit matrix metalloproteases without cytotoxicity. J Dent Res 94:1085

34. Popovtzer R, Agrawal A, Kotov NA, Popovtzer A, Balter J, Carey TE, Kopelman R (2008) Targeted gold nanoparticles enable molecular CT imaging of cancer. Nano Lett 8:4593

35. Wong RC, Tse HY (eds) (2009) Lateral flow immunoassay. Humana Press, Springer, New York

36. Wild DG (ed) (2013) In: The immunoassay handbook: theory and applications of ligand binding. ELISA and related techniques, Elsevier, Oxford

37. Bahadır EB, Sezgintürk MK (2016) Lateral flow assays: principles, designs and labels. TrAC—Trends Anal Chem 82:286–306

38. Andryukov BG (2020) Six decades of lateral flow immunoassay: from determining metabolic markers to diagnosing COVID-19. AIMS Microbiol 6:280–304

39. Sajid M, Kawde AN, Daud M (2015) Designs, formats and applications of lateral flow assay: a literature review. J Saudi Chem Soc 19:689–705

40. Nutting J, Nuttall JL (1977) The malleability of gold. An explanation of its unique mode of deformation. Gold Bull 10(2)

41. Koczula KM, Gallotta A (2016) Lateral flow assays. Essays Biochem 60:111–120

Chapter 2
Dental Gold Alloys

2.1 Classification and Selection

Gold has been a dental restorative material for more than 4000 years; from early dental applications based on aesthetics, then gold wire to bind teeth used by early Phoenicians, and, finally, the Etruscans and the Romans, who introduced the art of making fixed bridges from gold strips. These techniques were lost during the Middle Ages. Rediscovery in a modified form came to fruition in the middle of the nineteenth century by the introduction of gold alloys [1].

Pure gold is a well-known restorative material with an exceptional combination of properties: Tarnish resistance, ductility with work hardening and the ability to be cold-welded by pressure alone. It is usable as a direct filling material in the form of foil or powder. Unfortunately, this process was more or less abandoned, due to its inappropriate price and additional insertion skills of restorative elements. However, pure gold, even though not very strong, is used for inlays, i.e. for crowns, bridges and removable partial dentures. For more advanced use, gold alloys with strength advantages are necessary. In that sense, hardening and strengthening must be done using suitable alloying elements [2–4].

Gold alloys have been used in Dentistry, not only because their gold colour is preferred, but also because they have extremely high chemical stability in the mouth, plus several desirable mechanical properties, such as high strength, ductility and elasticity when it is used in an alloy. Gold alloys are formed by the use of macro and micro alloying elements. In the molten state these metals dissolve to various degrees in one another, allowing them to form Au alloys in the solid state. On the other hand, the solid solubility depends on the relative sizes of the individual atom species, the crystal structure formed by the pure metal components, the valences of the components and their reactivity. Minor changes in the chemical composition, and, thus, the temperature of the technological process, lead to changes in the structure/microstructure or in the phase composition, which can improve the final properties of gold alloys [4, 5].

© The Author(s), under exclusive license to Springer Nature Switzerland AG 2022   13
R. Rudolf et al., *Dental Gold Alloys and Gold Nanoparticles for Biomedical Applications*,
SpringerBriefs in Materials,
https://doi.org/10.1007/978-3-030-98746-6_2

As we know, gold is soft and can be strengthened through alloying. The conventional alloying elements typically used include silver (Ag), copper (Cu), nickel (Ni), platinum (Pt), palladium (Pd), manganese (Mn) and chromium (Cr). It can be stated that Ag, Cu and Ni are the main strengthening elements, and Mn and Cr are used mainly as resistance-sensitive elements. Gold can also be strengthened by microalloying. Of the possible microalloying elements, rare earth (RE) metals are very effective strengthening agents for both pure gold and gold alloys [6, 7].

The use of three—and more—components for the preparation of gold alloys in the production of final dental gold products is necessary, in order to obtain the appropriate clinical features. Due to the complexity of three-component alloys, which cannot be drawn in a plane, the most common starting point is the study of two-component alloys, Au-Ag, Ag-Cu and Au-Cu, as well as Au-Zn and Au-Pd. Diagrams are created in general on the basis of data generated by research from the following areas: Cooling curves, metallographic investigations, chemical tests, crystal structure analysis, information on mechanical, physical and other properties, etc. This leads to the construction of the basic and most important three-component alloys of the Au-Ag-Cu system, characterised by complete solubility in the liquid state, and fairly high solubility in the solid state. The lowest melting point is the triple eutecticum, at a temperature of 767 °C, composition of 14.0 at.% Au, 43.0 at. % Ag, 43.0 at. % Cu. The eutectic point of the triple Au-Ag-Cu diagram moves, by adding gold, to the angle rich in gold, starting from the eutectic point of the Ag-Cu state diagram. The two-phase minimum at the liquidus line of the Au-Cu state diagram, located at a temperature of 910 °C, is projected into a triple state diagram, and eventually turns into a monovariant eutecticum, which ends at the eutectic point of the double Ag-Cu state diagram, at a temperature of 780 °C [7–9].

Dental gold alloys (DGA) as part of noble dental alloys are usually incorporated in the American Dental Association (ADA) compositional classification system developed empirically by manufacturers in 1984 [10]. This system is based on the three groups, divided by the noble metal content, treating only gold, platinum, and palladium as noble metals, Table 2.1.

This classification [10] identifies alloys as high-noble, noble, or predominantly base-metal. The general assumption is that alloys with higher noble metal content will corrode less. The percentages used as boundaries between categories are arbitrary, however, and the correlation between the categories and corrosion is not perfect by any measure. Thus, although high-noble alloys will generally corrode less intraorally than noble or base-metal alloys, there are numerous exceptions.

Table 2.1 American Dental Association (ADA) compositional classification system for noble dental alloys [10]	Class	Composition
	High-noble	Au content ≥ 40 wt. % Noble metal content ≥ 60 wt. %
	Noble	Noble metal content ≥ 25 wt. %
	Predominantly base-metal	Noble metal content < 25 wt. %

High-noble alloys have at least 40 wt. % of gold and 60 wt. % of noble elements in their composition, placed in the three common subclasses:

(I) *Gold-platinum alloys* are usually used for full cast or metal-ceramic applications. They may contain zinc or silver as hardeners, and are often multiple-phase alloys. These alloys were developed initially as palladium-free alternatives in the early 1990s.

(II) *Gold-palladium alloys* may also be used for full cast or metal-ceramic restorations. These alloys may or may not contain silver, but almost always contain tin, indium, or gallium as oxide-forming elements to promote porcelain adherence. Gold-palladium alloys are commonly selected for metal-ceramic restorations.

(III) *Gold-copper-silver-palladium alloys* are used exclusively for full cast restorations. The solidus temperatures of these alloys are too low for metal-ceramic applications, and the copper and silver content is often problematic during ceramic application.

The properties of high-noble alloys are generally favourable for manipulation and clinical service, but none of these alloys have a high elastic modulus value. Many, but not all, of these alloys are single-phase; the gold-platinum-zinc systems are one exception, for example. When palladium or platinum contents are above 10 wt%, the solidus temperatures of the alloys are higher, and the alloys are white in colour. The corrosion of these alloys is generally low, but may be higher if multiple phases are present.

Noble alloys have no stipulated gold content, but must contain at least 25 wt. % noble metal. This is a very diverse group of alloys, with gold-, palladium-, and silver-based systems represented. There are three common subclasses of alloys in this class:

(I) *Gold-copper-silver-palladium alloys* are a lower-gold variation of the high-noble gold-copper-silver-palladium alloys. Generally, the silver or copper is increased to compensate for the gold content. These alloys are always single-phase.

(II) *Palladium-copper alloys* are used for full cast or metal-ceramic applications. These alloys commonly contain gallium, which lowers the liquidus temperature, can provide improved porcelain adherence and contributes to strength.

(III) *Palladium-silver (or silver-palladium) alloys* are very diverse, and range from systems with only 26 wt. % palladium and more than 60 wt. % silver to alloys with 60 to 70 wt% palladium and approximately 20 wt. % silver.

As a group, the noble alloys have moderately high solidus temperatures, which reflect the higher palladium content. The exception is Au-Cu-Ag-Pd noble alloys, which have solidus temperatures that are too low for use in metal-ceramic restorations. The noble alloys may be yellow or white in colour, but are more often white, reflecting the high concentration of palladium in many formulations. Noble alloys are generally very strong, with good hardness and moderate percentage elongation (10–20%). The elastic moduli of palladium-copper-gallium and palladium-silver alloys are significantly higher than those of high-noble alloys (because of the former's higher palladium content), but only about 60% of the elastic moduli of base-metal alloys. Corrosion of the noble alloys is variable; it depends on the microstructure and the presence of corrosion-prone microstructural phases, such as silver and copper. Alloys with high hardness (>250 kg/mm^2) may be difficult to cut, shape, and/or polish.

Base-metal alloys contain less than 25 wt. % noble metal, but, in practice, most contain no noble metal. There are three subclasses:

(I) *Nickel–chromium*,
(II) *Cobalt-chromium* and
(III) *Titanium alloys*.

A review of the approximate composition and properties of the usual high-noble and noble gold alloys is displayed in Table 2.2 [9].

Table 2.2 Approximate composition and properties of high-noble and noble gold alloys[*]

Subclass	Approximate composition (major elements, wt. %)	Elastic modulus (GPa)	Vickers hardness (kg/mm2)	Yield strength [†] (MPa, 0.2% offset)	CTE ($\times 10^{-6}$ °C)
High-noble gold alloys					
Au-Pt	Au 85; Pt 12; Zn 1 (Ag)	65–96	165–210	360–580	14.5
Au–Pd	Au 52; Pd 38; In 8.5 (Ag)	105	280	385	14.3
Au–Pd	Au 72; Cu 10; Ag 14; Pd 3	100	210	450	NA
Noble gold alloys					
Au-Cu-Ag-Pd	Au 45; Cu 15; Ag 25; Pd 5	100	250	690	NA

[*] All properties are in the hardened condition, when applicable. Adapted from [9]
CTE = Coefficient of thermal expansion.
[†] In tension mode.
[‡] Elements in parentheses are present in some formulations.

The widespread use of gold alloys in Dentistry is shown clearly in Table 2.3 [2]. Some representative indications are shown in Fig. 2.1.

The mechanical property requirements proposed in EN ISO 1562 for casting gold alloys with a minimum gold content of 65% (m/m) and minimum content of gold and PGM's of 75% (m/m), and for dental casting alloys with noble metal content of at least 25% (m/m) but less than 75% (m/m) according to EN ISO 8891, are presented in Table 2.4.

Besides the inevitable cost, the selection of dental gold alloys with the appropriate physical, chemical and biological properties for a specific clinical situation should always be the first priority. In that sense, several properties of gold alloys will be discussed that are critical to the clinical performance.

Table 2.3 Compositions and indications of dental gold alloys [2]

Alloy Group	Gold content % (m/m)	Others % (m/m)	ISO EN Standards	Indications
1	99.9–99.99	–	–	A, M
2	97.9–98.3	1.7 Ti, Ir, Rh, Nb	9693	C, D, E
3	75–90	10–20 PGMs, In Sn, Fe, Re, Ag, Cu, Zn, Ta, Ti, Mn	9693	C, E
4	60–75	approx. 10 PGMs, 10–25 Ag, Cu, In, Sn, Zn	1562 9693	B, C, D
5	65–75	5–10 PGMs, 5–20 Ag, Cu, Zn	1562	B, D, E, F, G, H, L
6	60	40 PGMs	–	H, I, N
7	40–60 40 - 60	5–10 Pd, 10–30 Ag, Pt, Cu, Zn, Ir	8891	B, D, L
8	40–60	20–40 Pd, 0–20 Ag, In, Sn, Cu, Ga, Ir, Re, Ru	9693	C
9	approx. 15	approx. 50 Pd, approx. 20 Ag, In, Sn, Ga, Ir, Re, Ru	9693	C
10	2–12	30–60 Ag, 20–45 PGMs, Cu, Zn, Ir,	8891	B, F, L
11	0.1–2	50–60 Pd, 25–40 Ag, In, Sn, Ga, Ir, Re, Ru	9693	C
12	0.1–5	70–80 Pd, Sn, Cu, Co, Ga, Ir, Re, Ru, Pt	9693	C
13	10–80	0–20 PGMs, 0–70 Ag, Cu, Sn, In, Zn, Co, Ni, Mn	9333	K

A—Electroforming, **B**—Inlays, Onlays, Fixed Crown- and Bridgework as Full Cast and for Polymer Veneers, **C**—Fixed Crown- and Bridgework for Porcelain Veneers, **D**—Telescopic Crowns and Milling Work for Removable Dentures, **E**—CAD/CAM, **F**—Orthodontic Wire,—G Clasp Wire, **H**—Prefabricated Attachments, **I**—Root Canal Posts, **K** Solders, **L**—Partial Denture Framework, Cast Attacments, Plates, Connecting Bars, Saddles and Splints, **M**—Gold Foil for Direct Fillings, **N**—Cast-On Technique and **PGMs** = Platinum Group Metals

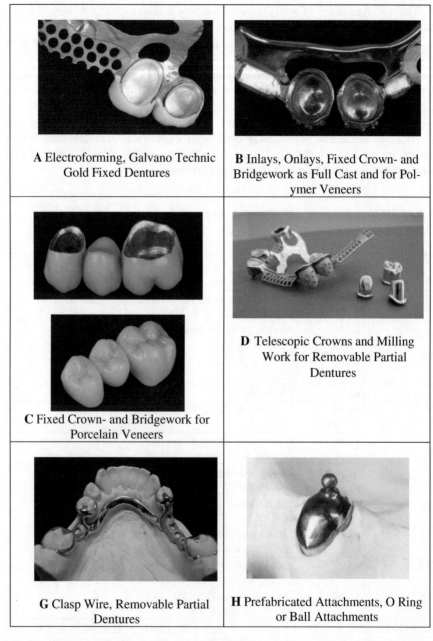

A Electroforming, Galvano Technic Gold Fixed Dentures

B Inlays, Onlays, Fixed Crown- and Bridgework as Full Cast and for Polymer Veneers

C Fixed Crown- and Bridgework for Porcelain Veneers

D Telescopic Crowns and Milling Work for Removable Partial Dentures

G Clasp Wire, Removable Partial Dentures

H Prefabricated Attachments, O Ring or Ball Attachments

Fig. 2.1 Representative indications of dental gold alloys (gratitude to the Dental Technician Mrs. Mila Simonović/Dental Lab Wisil M, Belgrade, Serbia)

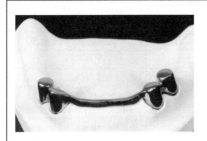

H Prefabricated Attachments, Bar Attachment

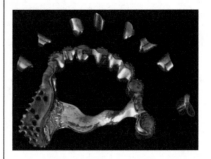

L Partial Denture Framework, Cast Attachments, Plates, Connecting Bars, Saddles and Splints

N Cast-On Technique

Removable Partial Denture Retained by a Telescopic Bridge with Composite Aesthetic Veneers Fused to External Crowns

Fig. 2.1 (continued)

Phase structure (microstructure with the grain structure of the alloy). Alloys can be either single phase (S), with the same composition throughout, or multiple phase (M), with areas of composition that differ by microstructural location. When elements are completely mutually soluble in the solid state, then the alloy is single-phase (for example, with gold and palladium or copper). If some elements are not soluble in one another, then the alloy is multiple-phase (e.g. gold and platinum).

The corrosion, strength and etching characteristics of alloys depend on their phase structure. Mainly, multiple phase alloys are inclined to higher corrosion rates than single-phase alloys. That is the consequence of the galvanic effects between the microscopic areas of different composition. On the other hand, multiple phases allow etching for bonding [11–13].

Table 2.4 Mechanical property requirements for dental casting gold alloys (Adapted from EN ISO 1562 and EN ISO 8891)

Type	Descriptor	Yield Strength (MPa)	Elongation (%)	Examples of Applications
1	Low	80	18	Inlays
2	Medium	180	10	Inlays and onlays
3	Hard	270	5	Onlays, thin cast backings, pontics, full crowns, saddles
4	Extra hard	360	3	Saddles, bars, clasps, crowns, bridges, and partial denture framework

Type 1: low-strength – for castings subject to very slight stress, e.g. inlays
Type 2: medium-strength – for castings subject to moderate stress, e.g. inlays and onlays
Type 3: high-strength – for castings subject to very high stress, e.g. onlays, thin cast backings, pontics, full crowns and saddles
Type 4: extra-high-strength – for castings subject to very high stress and thin cross-section, e.g. saddles, bars, clasps, thimbles, unit castings and partial frameworks

The effect of phase structure on strength is more complex. In some cases, multiple phases strengthen an alloy greatly, but in others they weaken the alloy. Strengthening depends on the nature of the second phase (particularly its ductility), its composition and dispersion throughout the other phases. Single-phase alloys are almost always easier to manipulate in the laboratory, have more consistent properties and are less technique-sensitive. It is for these reasons that manufacturers have preferred to sell single-phase alloys. The phase structure of an alloy is not visible without substantial magnification (alloys generally are viewed under a Scanning Electron Microscope) [6].

Small amounts of secondary phases can often be found at the grain boundaries of the major matrix phase [5].

Grain size. The grains with grain boundaries (see Tables 2.13 and 2.14) are important for the physical properties of the alloys, because the size and shape of the grains affect other clinical properties. For instance, small grains improve the elongation and tensile strength of cast gold alloys, but do not influence their hardness or yield strength. Grain sizes vary from 10 to 1000 μm. In general, a grain size of 30 μm or less is acceptable in Au dental alloys [11].

Dendritic structures (roughly analogous to grain structure) may be very large in base-metal alloys, where the size of a single grain can approach the diameter of a removable partial denture framework clasp.

Mechanical properties (yield strength, hardness and elastic modulus) are important for a good clinical performance of dental gold alloys. E.g. A permanent distortion of the dental restoration will occur when the tensile yield strength is exceeded. Usually, the position for this type of failure is between the pontics in a multiple unit fixed partial denture. Alloys with tensile yield strengths above 300 MPa are strong enough to resist permanent intraoral deformation in most clinical situations. The hardness of the dental alloy must be enough to resist occlusal forces but not wear opposing teeth. Generally, dental alloys with a Vickers hardness of less than 125 kg/mm^2 are susceptible to wear, and dental alloys that are harder than 340 kg/mm^2 (hardness of the enamel) are at risk of wearing opposing teeth. The elastic moduli for prosthodontic alloys need to be high, so that the prosthesis can resist flexure, especially in metal-ceramic restorations where any flexure will cause fracture of the porcelain. The elastic moduli of most gold- or palladium-based dental alloys range from 90 to 120 GPa (based on manufacturers' alloy properties charts). Such elastic moduli are sufficient in most clinical situations, but may not be adequate for long-span metal-ceramic restorations or removable partial dentures. In the latter cases, nickel- or cobalt-based alloys, which have a moduli of 180–230 GPa (Table 2.4), may be more appropriate [3, 4].

The colour of the dental alloys has been the focus of Dentists and patients for many years. Historically, yellow-coloured dental alloys have been associated with high gold content, high cost, high social value and good clinical performance. Similarly, white- (silver-) coloured dental alloys have been associated with corrosion, less social affluence and low cost. However, in today's alloy market, colour is useless in making any judgements about composition, cost, or clinical performance. There are many examples of high-gold dental alloys that are white in colour. In fact, any dental alloy that contains greater than 10 wt. % palladium will be white, regardless of the gold content. Similarly, dental alloys exist that have a yellow colour, even with no gold present (Pd-In-Ag, for example). As described previously, the cost of dental alloys is not necessarily related to the gold content, but can be influenced by the cost of other non-yellow elements, such as palladium, platinum and silver. The clinical performance of dental alloys is related to almost every physical property except colour. Thus, the Dentist should never make a clinical judgment about the dental alloy based on its colour. The patient's association of social value with yellow coloured dental alloys may play a part in the clinical decision, but it should never be the sole basis for dental alloy selection [9, 14, 15].

Corrosion of the gold dental alloys is of central importance to the success of a prosthesis. For metals and alloys, corrosion is always accompanied by a release of elements and a flow of current. Virtually every alloy known will corrode to some extent intraorally, but alloys vary significantly in this regard. Corrosion can lead to poor aesthetics, compromise of physical properties, or increased biological irritation. Corrosion can be measured either as the current flow or the amounts of released elements [12, 13]. Both methods are used commonly, but elemental release is probably more relevant to any adverse biological effects that corrosion might have. Illustrative examples of mass release from various dental gold casting alloys are shown in Table 2.5 [9].

Corrosion is complex and impossible to predict based simply on the composition of the dental alloy. The presence of multiple phases or high percentages of non-noble elements does, however, increase the risk of corrosion. In Dental Metallurgy, seven elements are recognised as noble: Gold, platinum, palladium, iridium, rhodium, osmium, and ruthenium. Corrosion of dental alloys may be clinically visible if it is severe, but, more often, the release of elements continues for months or years at low levels and is not visible to the eye. Corrosion is clearly related to biocompatibility,

Table 2.5 Release of mass from various dental gold casting alloys (μg/cm^2/day)

Type of alloy	ADA classification	Phases	Average mass released* (μ/cm^2/day)
Au-Pt	High noble	M	0.071
Au–Pd	High noble	S	0.005
Au-Cu-Ag	High noble	S	0.152
Au–Ag-Cu	Noble	S	0.1 84

S—Single; M—multiple.
* An average dental crown would have 2–3 cm^2 of surface area.
(Based on 10 months of study, data adapted from [9, 19]

but the relationships between them are complex and difficult to predict. Currently, the only way to know the biological effects of dental alloys is to test them for biocompatibility in vitro and in vivo [16–18].

Biocompatibility. Biocompatible materials are those materials that are used in contact with cells, tissues or body fluids of the human body without the presence of any adverse reactions. They are most often used to replace or upgrade the structural components of the human body, in order to compensate for damage caused by ageing, disease or accidents. In general, biocompatibility could be presented as the harmonious relationship between the host tissue, the material and the function of the material in the patient's body. Biocompatibility is a descriptive term denoting the ability of a material (*here, Dental Gold Alloys, DGA*) to behave appropriately within the organism in which it is incorporated, i.e. to perform a certain required function in the human body without causing an unwanted response of the host tissue [16]. Wintermantel et al. [17] extended this definition by emphasising that there is a difference between the surface and structural compatibility of implants. Surface compatibility means the chemical, biological and physical suitability of the implant surface for use in the host tissue, while structural compatibility represents the optimal adaptation of the implant to the mechanical behaviour of the host tissue. Therefore, structural compatibility is, on the one hand, related closely to the mechanical characteristics of the implant material, such as modulus of elasticity and strength, and, on the other hand, to the designed implant shape and optimal load transfer along the implant and tissue separation line.

The basic requirement for the use of DGA in Dentistry is to prove their safety, i.e. to prove that this Au material does not cause local or systemic cytotoxicity, irritation and allergy, that it is not mutagenic and carcinogenic. No material is completely biologically inert, so the term "degree of biocompatibility" is often used in the literature. During the contact of the DGA with living tissues, numerous complex interactions occur, which are collectively called the biological response to the applied Au material. The type of biological response depends on: (a) The types of Au materials, (b) The length of contact of the Au material with the tissue, (c) The host organism in which the Au material is incorporated, and (d) the Au material's functions.

Both the Au material and the host change over time, making biocompatibility a dynamic process. Generally, materials used to make medical implants, regardless of the type and nature of implantation, must meet certain criteria and possess appropriate properties, such as: Biocompatibility, bioadhesiveness, bioinertness and biofunctionality [18–20].

Ageing. Practically, dentists and doctors ask engineers for explanations for the phenomena that take place in the DGAs. These embedding materials dissolve, release ions, or otherwise act on the surrounding tissue—the host, causing or not reactions in the surrounding tissue. However, laboratory procedures for shaping dental restorations, as well as the surrounding tissue of the oral cavity, affect the embedding material, causing certain changes, the *ageing* of the dental gold alloys.

Traditional high-gold dental alloys used to make telescopes and faceted crowns are reinforced in the cast state by adding other elements (silver, copper, platinum, palladium and zinc) to a surface-centred cubic crystal lattice (fcc) of the solid gold solution [21, 22]. Such high Au content dental alloys should contain a sufficient % of copper, in order to ensure an increase in hardness and strength, and in order to form ordered regions of AuCu by heat treatment—annealing at elevated temperatures [23, 24]. There are two arranged AuCu super-lattices: AuCu II at higher temperatures and AuCu I at lower temperatures [25]. The ordered $AuCu_3$ structure is important for the ageing of materials of popular alloys with lower gold content [26], and its role in the ageing of the material has also been studied and confirmed by Transmission Electron Microscopy (TEM) [27, 28]. High-resolution Transmission Electron Microscopy has shown that hardening and strengthening of old gold dental alloys with high gold content is due to the interaction of dislocations with elastic stress fields at the interface between the arranged AuCu I planes, which have a surface-centred tetragonal (fct) structure, and a disordered fcc matrix gold solution. The analogous dislocation of the interaction with the elastic stress fields at the interface with the arranged $AuCu_3$ regions [27–29] is considered to be responsible for the ageing of alloys with reduced gold content.

Dental manufacturers provide guidance on appropriate heat treatment procedures to increase the strength and hardness of a specific gold dental alloy. Unfortunately, these procedures are time consuming, and most often do not apply to spills.

During the usual procedures in dental laboratories [25], cast restorations made of gold dental alloys are immersed in water when the casting loses its red hot appearance, which leaves the casting in a softened state. This is the condition in which the alloy is after the final correction of the restoration, when it is cemented to the abutment teeth. In order to ensure the highest possible hardness and strength of gold dental alloys with Cu, the restoration is heat-treated—annealed, usually at a temperature between 200 and 450 °C for 15–30 min, depending on the product and the manufacturer's instructions. It is recommended that the dental alloy be completely softened before it is annealed to harden. It is usually annealed for 10 min at a temperature of 700 °C, followed by forced cooling, immersion in water, in order to prevent the formation of a homogeneous coarse-grained structure that would occur during slow cooling to room temperature. Annealing of the casting converts the gold dental alloy into a single-phase alloy with a disordered fcc solid solution.

New high-carat gold dental alloys that can age, and have better biocompatibility, are alloys to which a small amount of titanium has been added. The relationship between isothermal ageing and the phase transformation of Au-1.6 wt. % Ti can be attributed to the continuous precipitation of the Au_4Ti ordered phase in a supersaturated solid α matrix solution [30].

***Ageing of dental gold alloys* in vivo.** The changes that occur during the ageing of gold dental alloys in vivo, despite the enormous advances in science, have not been explained fully. As the facts are explained, for example, removing a gold alloy bridge after 20 or more years of wearing does not show signs of material fatigue (ageing), while, on the other hand, a modern metal-ceramic bridge shows serious signs of material fatigue during the "warranty" period. The following lines therefore describe some important principles of ageing of building dental alloys in vivo.

All conventional precious and base dental alloys contain multicomponent elements. According to the Gibbs phase rule [3], these alloys contain multiple phases at intraoral and room temperature under equilibrium conditions. Even if complex multi-component phase diagrams are available for dental alloys, they do not provide complete information. Due to insufficient time for diffusion of the elements, which would maintain the equilibrium phase composition in conditions of rapid hardening in the dental laboratory, the microstructure of cast dental alloys inevitably contains micro-segregation of elements, as well as non-equilibrium phases. These phenomena are visible when the microstructures of gold-based dental alloys and alloys with high palladium content are compared in the cast state, and after heat treatment that simulates the firing cycles of ceramics, where microstructural homogenisation is observed [31]. Conventional based dental alloys for metal-ceramic restorations [32] and partial skeletal prostheses [33, 34] in the cast state have a dendritic microstructure. Studies have shown that thermal treatment at elevated temperatures of base dental alloys for partial skeletal prostheses caused microstructural changes with a reduction in strength [35, 36].

A repeated and cast dental alloy in the production of fixed dental restorations and metal skeletons of partial dentures leads to accelerated ageing of the alloy. Changes caused by the combustion of micro-constituents (deoxidants) and binding of some elements (if poured on an open flame or in an induction apparatus in air), affect the microstructure of the dental alloy, and are responsible for changes in the mechanical and other characteristics of the alloy [37–39]. Altered physical and mechanical characteristics of the dental alloy affect the environment in which dental restorations are placed adversely, which, in turn, leads to adverse reactions that may affect the tissue response in the immediate environment.

Although opinions in the professional literature are divided on whether it is good to melt and cast one dental alloy more than once, we believe that obvious changes in the mechanical characteristics of the dental alloy and increasing patient sensitisation to metals suggest that manufacturers' instructions on melting and casting conditions must be followed. It can be said generally that fixed dental restorations should be made from dental alloys in their original form [40].

Porcelain-bonding properties. Porcelain-fused-to-metal restorations have become a widespread technology in the last 20–30 years. Gold dental alloys with special properties, including high sag resistance and compatible coefficients of thermal expansion to porcelains, have been developed, fulfilling the necessary requirements for fixed crowns and long span bridges with porcelain veneers. The EN ISO 9693:2000 Standard describes the requirements for these alloys, which are not divided into different types, nor have a prescribed gold content. This Standard is valid for noble and base metal dental alloys, as well as for the suitable porcelains [4, 41, 42].

The main specific physical requirements for those dental alloys used for the porcelain fused to metal (PFM) technique are: (a) A melting range starting at no less than 1100 °C, (b) A coefficient of thermal expansion closely matched to that of the high-fusing-point dental porcelain (960–980 °C) developed for the PFM technique—the values for these coefficients should stay within the ranges 12.7–14.8 × 10^{-6} K^{-1} for the alloys, and 10.8–14.6 × 10^{-6} K^{-1} for the porcelain, (c) Minimal creep or sag when firing the porcelain, (d) Good wetting of alloy by the porcelain, (e) A tensile strength about 600 N/mm^2 and yield strength of about 550 N/mm^2, (f) Hardness of about 180 HV and (g) and elongation of about 10%. The general consensus is that the dental alloy should have a higher coefficient of thermal expansion than the porcelain (a positive expansion coefficient mismatch), in order to produce compressive stress in the porcelain when cooling [43]. On the other hand, gold alloy for casting should have a melting range as low as possible, whilst an Au alloy for porcelain veneers should have a solidus temperature around 100 °C higher than the firing temperature of the used porcelain, in order to ensure that the cast framework does not sag during firing [6].

For a porcelain restoration, several additional properties of a gold dental alloy are important for clinical performance. The colour and thickness of the alloy oxide must be considered. High-gold dental alloys have a relatively light-coloured oxide that is easier to mask with opaque porcelain, whereas most silver-, nickel-, and cobalt-based dental alloys have darker, grey oxides that require thicker layers of opaque porcelain to mask. If these oxides are not masked completely, they will impart a lower value to the porcelain shade. In general, research has shown that thicker oxides increase the risk of metal-ceramic bonding failure.

Fracture tends to occur because oxides are brittle and weaker than either the porcelain or the alloy. Furthermore, stress from occlusal loads is often concentrated in the oxide layer. The thickest oxide layers occur in nickel- and cobalt-based dental alloys, because these alloys contain elements that form oxides easily during the initial oxidation step (historically and incorrectly termed "degassing") prior to firing of the opaque porcelain. Gold and palladium form oxides much more sparingly because of their noble character; alloys based on these metals therefore require the addition of tin, gallium, indium, or other trace elements, to promote oxide formation. Even with such additions, oxides in gold- and palladium-based dental alloys are thinner [4].

Caution must be exercised in reusing gold-based dental alloys for metal-ceramic restorations, because the oxide-forming elements may be depleted from the first casting procedure. In that situation, the oxide layer would be inadequate for reliable bonding of the porcelain. The relative expansion between the metal and ceramic is of prime importance in porcelain bonding. Both alloys and porcelain expand when heated and contract when cooled. If porcelain and an alloy bond together at a high temperature (the sintering or firing temperature of the porcelain), then the relative contraction rates of these two materials will be important as the bi-material bond cools to room temperature. If the porcelain contracts less than the dental alloy as cooling progresses, then the porcelain will have residual compressive stress at room temperature. If the porcelain contracts more than the alloy, then the porcelain will have residual tensile stress at room temperature.

Because porcelain is a brittle material, and, thus, subject to failure by crack propagation, it does not tolerate tensile stresses well. Thus, the metal-ceramic bond must minimize the residual tensile stresses in the porcelain. It is best to select a porcelain with a coefficient of thermal expansion (and contraction) that is less than that of the alloy. Most dental alloys have coefficients of thermal expansion between 13.5 and $17.0 \times 10^{-6} \text{ K}^{-1}$. Traditional ceramics have coefficients of $13.0–14.0 \times 10^{-6} \text{ K}^{-1}$, but newer ceramics may vary substantially from this range. It is, therefore, critical to consult the alloy manufacturer when selecting a porcelain for a given dental alloy. It is important to note that the coefficient of thermal expansion for the porcelain cannot be too much smaller than the alloy, or the porcelain-metal bond will fail as a result of compressive stresses. Generally, a $0.5 \times 10^{-6} \text{ K}^{-1}$ difference in coefficients is desirable.

Dental alloys for metal-ceramic restorations must have melting temperatures that are compatible with the peak firing or sintering temperature of the porcelain. Because they are mixtures of elements, alloys have melting ranges rather than single melting point temperatures [9] (Table 2.6). Each alloy has a lower temperature (called the solidus temperature) at which melting begins, and an upper temperature (called the liquidus temperature) at which the entire alloy is melted. During porcelain firing or soldering operations, it is critical to stay below the solidus temperature of the alloy. Most laboratory technicians recommend using a porcelain that sinters at least 50 °C below the alloy solidus temperature (porcelain fuses from 870 to 1370 °C) [35] to prevent distortion of the alloy substructure at high temperatures. If a restoration is to be soldered after the application of the porcelain, then the solder must have a liquidus temperature at least 50 °C below that of the porcelain sintering temperature and the solidus temperature of the dental alloy.

Table 2.6 Solidus and liquidus temperatures of current common classes of prosthodontic alloys[*]

Alloy type	ADA Classification	Solidus temperature (°C)	Liquidus temperature (°C)
Au-Pt	High-noble	1060	1140
Au–Pd	High-noble	1160	1260
Au-Cu-Ag-Pd	High-noble	905	960
Au-Cu-Ag-Pd	Noble	880	930

[*] Information adapted from [9]

2.2 Examples of Expertise

Today, a number of new alternative materials are emerging with intensive development in the field of Dentistry and Dental Materials. It is no coincidence that Au based dental alloys are set as the Standard for assessing the quality of new materials. If we evaluate dental alloys from the aspect of the most important selection criterion (biocompatibility), noble single-phase alloys with high Au content are in the first place. Precious single-phase alloys with a high gold content show the least tendency to dissolve atoms in the oral cavity. Experience from practice and laboratory tests show that, also, in the case when, in addition to biocompatibility, the aesthetic appearance of a dental restoration is placed in the first place, the best choice is a noble dental alloy. Finally, if the other necessary criteria that must be met by dental materials are taken into account (durability, appropriate mechanical and physical characteristics, functionality, manufacturing and processing technology, aesthetics, durability of investment, required laboratory equipment), precious dental alloys have proven to be the optimal choice.

2.2.1 Mechanical Properties of DGAs in Terms of Tensile Testing and Hardness

Tensile testing. Eleven different DGAs of the multiphase system were prepared according to ISO 9693:2000, and tested extensively with a universal testing machine (Z010; Zwick Roell, Germany). The compositions of the tested DGAs and their

Table 2.7 Composition of the investigated DGAs

Comercial name	Composition (wt. %)	Symbols
Auropal SE	Au 2.0%; Pd 25.0%; Ag 64.0%; Cu 8.0%; Zn < 1%	Aac[*], Ah[*]
Auropal S	Au 10.5%; Pd 21.0%; Ag 58.2%; Cu 9.3%; Zn < 1%	Bac, Bh
Midor SE	Au 40.0%; Pd 4.0%; Ag 47.0%; Cu 7.5%; Zn, Ir, In < 1%	Cac, Ch
Midor S	Au 46.0%; Pd 6.0%; Ag 39.5%; Cu 7.5%; Zn, Ir < 1%	Dac, Dh
Dentor S	Au 75.5%; Pd 1.2%; Pt 4.4%; Ag 11.0%; Cu 6.7%; Zn 1.2%	Eac, Eh
Dentor N	Au 77.0%; Pd 2.6%; Pt 1.4%; Ag 12.0%; Cu 6.0%; Zn < 1%	Fac, Fh
Dentor BIO	Au 78.3%; Pt 5.0%; Ag 10.0%; Cu 6.7%	Gac, Gh
Aurodent BIO	Au 83.3%; Ag 10.0%; Cu 6.7%	Hac, Hh
Aurodent 22	Au 91.6%; Ag 5.0%; Cu 3.4%	Iac, Ih
Aurodent 20	Au 83.3%; Ag 10.0%; Cu 5.9%; Zn < 1%	Jac, Jh
Aurodent 18	Au 75.0%; Ag 10.0%; Cu 15%	Kac, Kh

[*] ac = as cast, h = hardening

abbreviations used in the rest of the text are shown in Table 2.7. The gold content was increased from 2.0 wt. % in the alloys Aac, Ah to 91.6 wt. % in Iac, Ih.

The tensile strength and toughness of DGAs depend mainly on the microstructure, which is defined systematically through its most important aspects: Creation and production, definition and analysis, and through the relationship to the macroscopic properties of the materials [21]. The microstructure of DGAs can be considered not only as an arrangement of crystal and phase boundaries, but also as a series of other defects and characteristics of the crystal lattice that extend practically all the way to the atomic structure. Therefore, the microstructure is defined as the totality of all crystal defects that can be observed, according to their type, number, distribution, size and shape [22]. A comprehensive definition of the microstructure is particularly important for determination of the structure and the final properties of DGAs. The fact is that the elements of the microstructure must be interpreted in such a way that they are compatible with the real process of formation, such as casting or solid-state reactions in DGAs (by annealing, hardening, or softening). On the other hand, the elements of the microstructure must be suitable for creating representation and quantification of relationships with macroscopic properties that allow a complete and quantitative description of the physical-metallurgical bases, such as: Dislocation mechanics, crystal lattice mechanics, the characteristics of two- and multiphase microstructures, fracture mechanics, recrystallisation, phase boundary and crystal structures, solid state reactions, a description of the texture and anisotropy relations, etc.

The grains in DGAs are formed randomly during growth with the same crystal structure, and are independent of the external form. Grain arrangement, including crystallite boundaries and other defects, is referred to as the microstructure.

The microstructure of DGAs is reflected in various forms of individual crystals and in their crystal structure and chemical composition. However, the microstructure should be understood not only by the eye and microscope noticeable interdependent configuration of solid areas separated by incoherent boundary surfaces from each other, but also the type, shape, size, density and distribution of crystal lattice defects.

Figures 2.2, 2.3 and 2.4, under (a), show the typical starting microstructures of the tested DGAs obtained with a Scanning Electron Microscope (SEM), Sirion 400NC (FEI, USA), located in the Faculty of Mechanical Engineering, University of Maribor. The revealed microstructures are multiphase, with grains of different sizes, and with the presence of some defects (borders, etc.) without a high porosity content. In the usual procedure in a dental laboratory, castings from DGAs are quenched with water after solidification. This results in the formation of a disordered substitution solid solution and leaves the DGA in a soft state, which is desirable, as the dentist usually also makes adjustments to the tooth structure (comparison with the patient's jaws). If the gold dental cast is in a hardened state, additional softening can be performed by heating at 700 °C for 15 min and quenching with water. The established quenching process of Au dental alloys is carried out at 350 °C for 15 min and with air cooling. This heat treatment results in the formation of ordered regions of AuCu or $AuCu_3$ in a disordered matrix of alloys with a high or lower gold content [28].

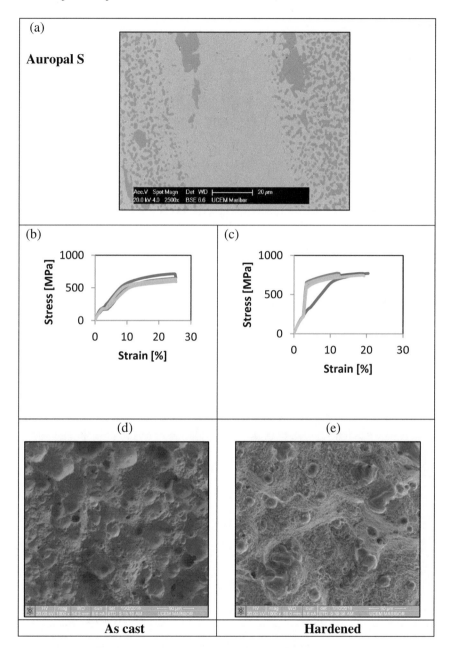

Fig. 2.2 **a** SEM micrograph of the investigated **Auropal S** surface after casting and cutting into dental plates (10 mm × 10 mm). Tensile strengths of: **b** As cast and **c** Hardened samples. Typical fracture surfaces of: **d** As cast and **e** Hardened samples

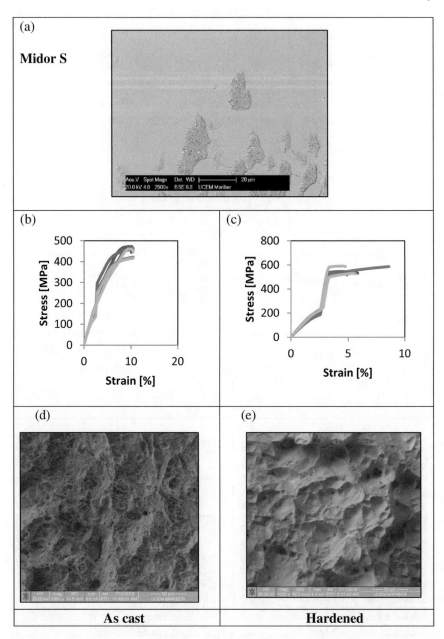

Fig. 2.3 a SEM micrograph of the investigated **Midor S** surface after casting and cutting into dental plates (10 mm × 10 mm). Tensile strengths of: **b** As cast and **c** Hardened samples. Typical fracture surfaces of: **d** As cast and **e** Hardened samples

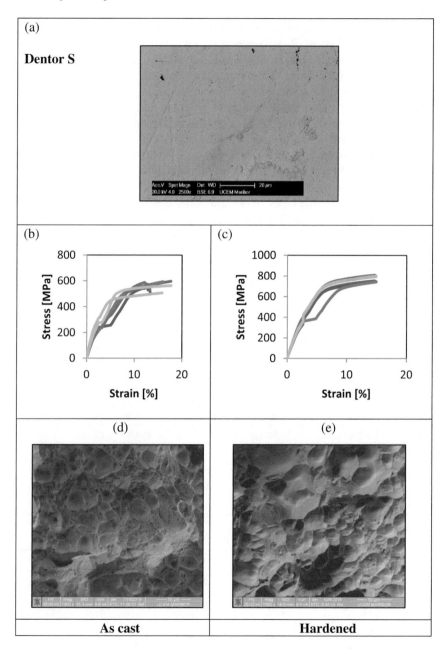

Fig. 2.4 **a** SEM micrograph of the investigated **Dentor S** surface after casting and cutting into dental plates (10 mm × 10 mm). Tensile strengths of: **b** As cast and **c** Hardened samples. Typical fracture surfaces of: **d** As cast and **e** Hardened samples

The mean values of 0.2% offset yield strength ($Rp_{0.2}$), Youngs modulus (E-Modulus), tensile strength (R_m), toughness (W up to break) elongation at maximum force (e Fmax.), and elongation after fraction (e Break), were calculated for each investigated gold dental alloy as the mean of at least six samples. The mean values of as cast and hardened gold dental alloys are presented in Tables 2.8 and 2.9. Although these alloys consist of a complex multiphase system of many different elements, the dependence of the chemical composition on their mechanical properties is evident.

The $Rp_{0.2}$, which is defined as the stress at which 0.2% plastic deformation occurs, varied greatly between the tested dental alloys. The highest $Rp_{0.2}$, with values of 602.29 ± 27.63 MPa and 746.46 ± 22.45 MPa, were calculated for the dental alloys Aac and Ah, with a composition of Au 2.0%; Pd 25.0%; Ag 64.0%; Cu 8.0%; Zn < 1% (see Table 2.7—first line). When the proportion of Au in the dental alloys Iac and Ih increased up to 91.6%, the $Rp_{0.2}$ decreased to 188.62 ± 9.11 MPa. The effect of the additives Pd, Cu and in Au alloys and their interaction on $Rp_{0.2}$ was also

Table 2.8 Mean values of mechanical properties for as cast DGAs

		Rp 0.2 (Mpa)	E-Modulus (GPa)	W up to break (Nmm)	Rm (MPa)	ε Fmax (%)	ε Break (%)
Aac	x[*]	602.29	104.615	11,578.4	795.17	22.48	23.36
	s[*]	27.63	1.81	2605.8	14.06	3.99	4.24
Bac	x	539.63	98.5	15,072.82	650.66	29.54	31.3
	s	20.28	2.43	1934.54	20.28	3.78	4.14
Cac	x	425.67	94.15	4392.65	557.12	9.51	10.43
	s	26.98	2.29	904.72	26.79	1.48	0.89
Dac	x	373.13	91.65	6972.25	454.9	12.72	13.52
	s	8.49	1.85	1710.09	21.47	3.36	3.16
Eac	x	453.81	88.2	6493.23	566.16	15.01	15.33
	s	24.51	2.11	1880.9	34.56	3.37	2.95
Fac	x	398.8	75.16	24,258.25	546.82	54.01	58.04
	s	8.57	1.9	1735.31	10.19	3.87	3.31
Gac	x	521.41	83.62	17,620.03	629.46	41.07	42.11
	s	20.72	1.87	4581.92	20.13	8.55	7.87
Hac	x	319.7	61.87	14,608.72	441.28	40.64	45.96
	s	10	1.94	3518.59	13.52	6.8	8.07
Iac	x	188.62	54.56	13,031.33	304.99	41.07	48.58
	s	9.11	2.55	1052.65	10.6	1.78	2.23
Jac	x	325.47	62.36	16,246.33	497.33	45.59	50.26
	s	10.04	1.84	2188.97	16.52	4.31	4.89
Kac	x	464.05	70.47	10,210.71	512.31	19.47	24.7
	s	25.68	1.68	1466.36	33.77	7.14	2.55

[*] x-mean value, s-Standard Deviation

Table 2.9 Mean values of mechanical properties for hardened DGAs

		Rp 0.2 (Mpa)	E-Modulus (GPa)	W up to break (Nmm)	Rm (MPa)	ε Fmax (%)	ε Break (%)
Ah	x[*]	746.46	108.85	15,671.02	910.66	25.46	25.92
	s[*]	22.45	2.47	1738.65	27.68	1.8	1.96
Bh	x	646.01	110.44	13,978.27	758.42	20.8	21.25
	s	13.26	1.41	2616.9	16.3	2.67	2.67
Ch	x	531.26	104.82	7122.61	649.2	12.76	13.24
	s	29.23	1.87	1636.74	22.75	2.68	2.71
Dh	x	495.51	100.28	5513.43	559.71	11.5	11.78
	s	21.43	2.62	1360.99	25.16	2.05	1.97
Eh	x	647.93	105.9	8938.16	780.94	16.2	16.52
	s	26.69	3.92	901.62	29.08	1.36	1.53
Fh	x	439.42	80.45	23,162.19	628.08	56.52	60.13
	s	15.71	4.23	2993.8	23.78	4.67	6.00
Gh	x	579.39	93.09	13,938.06	700.81	36.1	38.77
	s	15.39	2.96	1406.66	15.6	2.45	2.63
Hh	x	394.18	70.26	17,000.39	540.24	48.32	53.03
	s	19.21	2.07	1895.35	25.28	4.25	3.49
Ih	x	254.78	70.86	13,567.78	370.77	41.35	49.68
	s	9.31	1.81	911.46	13.98	1.54	1.48
Jh	x	422.87	65.59	16,015.13	581.31	45.63	49.34
	s	12.96	1.53	931.94	21.82	3.11	1.17
Kh	x	554.98	80.78	3051.25	637.12	6.41	9.69
	s	18.9	2.36	574.72	28.19	1.77	1.20

[*] x-mean value, s-Standard Deviation

significant. In principle, the higher content of Pd and Cu increases the $Rp_{0.2,}$ since both pure Pd and Cu have a high E- modulus. In the case of Gac and Gh alloys with Pt added, the $Rp_{0.2}$ remained above 500_{MPa}, despite the higher Au content. In most cases, hardening increased the $Rp_{0.2}$ by a further 10%. Hardening in an alloy system is probably due to the formation of an AuCu superlattice [30].

Figures 2.2, 2.3 and 2.4 under b and c, show the typical tensile strengths of the mechanically tested DGAs (as classified in Table 2.7).

In all samples from A to K, the rigidity or E-modulus of the DGAs was measured by the slope of the elastic region of the stress-strain graph. E-modulus is an essential design value when computing deflections of prostates by structural mechanics [44]. As cast dental gold alloys have their E-modulus ranged from 54.56 ± 2.55 GPa up to 104.615 ± 1.81 GPa. Under hardening, dental gold alloys have an E-modulus ranged from 65.59 ± 1.53 GPa up to 108.85 ± 2.47 GPa. The highest was in the case of the Bh alloy with a composition of Au 10.5%; Pd 21.0%; Ag 58.2%; Cu 9.3%; Zn <

1%, followed by the Ah alloy and Ch with compositions of Au 2.0%; Pd 25.0%; Ag 64.0%; Cu 8.0%; Zn < 1% and Au 40.0%; Pd 4.0%; Ag 47.0%; Cu 7.5%; Zn, Ir, In < 1%, respectively. A small amount of additives such as Pd, Cu and Pt improves the E-modulus, as well as the strength and bond strength [26]. With higher proportions of Au and Ag in multiphase alloys it is possible that their crystal structure changes, resulting in a lower E-modulus [27]. The ability of dental gold alloys to absorb energy and deform plastically without fracturing is called toughness. The highest toughness was calculated for the cast Fac and Fh alloys with a composition of Au 77.0%; Pd 2.6%; Pt 1.4%; Ag 12.0%; Cu 6.0%; Zn < 1%, followed by the Hh alloy with a composition of Au 83.3%; Ag 10.0%; Cu 6.7%.

The maximum load that a DGA can withstand without breaking when stretched is indicated by the tensile strength. As shown in Tables 2.8 and 2.9, the ability of the DGAs to resist induced stress without fracture depends on their composition. A significant increase in strength was found with increasing the Pd content and decreasing the Au content. The alloys Iac and Iah, with a composition of Au 91.6%; Ag 5.0%; Cu 3.4% had the lowest Rm at 304.99 ± 10.6 MPa, but one of the highest ε Fmax and ε Break, at 48.58 ± 2.23%. The highest elongation was observed in the alloys Fac and Fh, with an ε Break at 58 ± 04 3.31% and 60.13 ± 6.00%. Similarly, high Fmax and ε Break were found in the alloys Hh, Jh and Gh, with a high content of Au. Hardening leads in most cases to an increase of Rm, and, simultaneously, to a decrease of ε Fmax. These results are comparable to the known values of elastic modulus of other gold-based alloys having values between 75 and 110 GPa [28], while the elastic moduli of many base metal dental alloys are considerably greater. The DGAs must have sufficient strength for their application in dental practice. For fully cast dental structures, the strength requirements increase as the number of tooth surfaces to be replaced increases [29]. As a comparison, dental alloys for bridgework require higher strength than those for single crowns. Pins for metal-ceramic dentures are made in thin sections, and a sufficient modulus of elasticity or stiffness is required to prevent excessive elastic bending due to the working of functional forces, especially when used for replacement long-span dentures.

Standard ISO 22674, Metallic Materials for Fixed and Removable Restorations and Appliances, classifies metallic materials that are suitable for the fabrication of dental appliances and restorations, including metallic materials recommended for use either with or without a ceramic veneer, or recommended for both uses, and specifies their requirements. It further specifies requirements with respect to packaging and marking the products, and to the instructions to be supplied for the use of these materials.

According to the relevant International Standard ISO 22674, classified by 0.2% yield strength, elongation and modulus of elasticity, the DGAs studied in Tables 2.8 and 2.9 fall under Types 1 and 2.

Figures 2.2, 2.3 and 2.4 under (e) and (e), show typical fracture surfaces of the mechanically tested DGAs (as classified in Table 2.7). As can be seen from the micrographs of the fractures, more brittle fractures can be detected in Au dental alloys in the state "as cast", which is reflected in the perception of smaller elongations of the tensile tubes after breaking. A brittle fracture is characterised by a very low level of

energy expended for its formation, as well as the complete absence of plastic deformations. The main difference between tough and brittle fractures is that, with a tough
fracture, at spreading the central crack, there must be significant plastic deformation,
while, in the case of a brittle fracture, the crack propagation is not conditioned by
plastic deformation. A brittle fracture occurs, as a rule, inside the crystal, and spreads
along the plain crystallographic planes of individual grains of polycrystals called the
refractive plane. The occurrence of a brittle fracture is related to both the structural
structure of metals and operating conditions, primarily the operating temperature,
application rate load and existence or absence of inclusions. From an engineering
point of view, a brittle fracture is a fracture that flows with minimal plastic deformation, and is conditionally called a brittle fracture. In the case of fractures from the
hardened Au dental alloys group, the examination shows that they have the characteristics of a tough fracture with the presence of longer rivers. They are characteristic
elements of tough fractures, and indicate plastic deformation of the fracture surface.
A tough fracture is characterised by intense plastic deformation in all stages of the
fracture, and occurs at a strength well above the yield stress. It is not necessary for
a tough fracture to create and propagate a crack. Plastic processes lead easily to
breakage deformations, and tough fractures are transcrystalline because the crack
moves through the crystal grains [45].

Hardness. According to the Vickers method, the hardness of the investigated
DGAs (see Table2.7.) are presented in Table 2.10. for their as cast and hardened
states. The hardness HV was measured on the ZWICK 3212 testing machine.

Chemically pure gold, like most metals, has too little strength and hardness to
be used in such a form in Dentistry. Today, there are numerous principles of rein-
forcement—hardening of gold dental alloys, some of which are used to improve
mechanical properties. These include alloying, hardening, dispersion strengthening
processes, etc. The relationships between chemical composition, heat treatment,

Table 2.10 Mean values of hardness for as cast and hardened Au dental alloys	Commercial name	Mean HV ± Standard Deviation (As cast)	Mean HV ± Standard Deviation (Hardened)
	Auropal SE	125.7 ± 4.5	185.7 ± 10.5
	Auropal S	199.3 ± 4.0	213.2 ± 9.4
	Midor SE	152.2 ± 3.2	205.7 ± 3.1
	Midor S	160.5 ± 6.6	224.8 ± 7.7
	Dentor S	182.8 ± 3.4	208.5 ± 10.0
	Dentor N	112.7 ± 2.1	175.5 ± 7.7
	Dentor BIO	124.8 ± 1.3	166.2 ± 5.8
	Aurodent BIO	93.2 ± 1.8	173.1 ± 9.2
	Aurodent 22	52.2 ± 1.4	75.6 ± 5.9
	Aurodent 20	87.9 ± 2.3	149.7 ± 6.3
	Aurodent 18	148.6 ± 3.1	187.8 ± 12.3

mechanical properties, hardness, and microstructure for many commercial gold dental alloys for casting dental structures have been investigated by a number of authors [30, 46, 47]. The results suggest that at least two different hardening mechanisms must occur in these Au dental alloys [48]. Ternary alloys outside the two-phase area of the gold-silver-copper system are hardened by microstructure arrangement. Commercial gold dental alloys and those in the two-phase range can be partially cured by editing, but another curing mechanism has been observed. This mechanism is not due to the observed precipitates at the grain boundaries, but is related to the observed intragranular needle structures, which have a mismatch with the base matrix and usually significantly higher hardness.

2.2.2   Thermocycling

Exposure of building dental alloys to alternating high and low temperatures and loads is an attempt to bring the experimental conditions as close as possible to real clinical conditions. In this way, artificial ageing changes that are a consequence of ageing are noticed faster than would be possible in clinical conditions.

Most studies on artificial ageing include thermocycling [44] (97.2%), while the number of studies that include exercise in addition to thermocycling is significantly lower (8.1%). The final temperatures at which dental alloy samples are most often exposed during thermal cycling are 50 °C and 55 °C, and 0 °C and 68 °C [50], respectively. The exposure time of samples to high and low temperatures varies from a few seconds to several hours. The number of temperature cycles to which the samples are exposed during thermal cycling ranges from 100 to 2500, and most often this number is between 250 and 300 cycles [45].

The conditions under which mechanical loading is performed are also different. Most often, the tested material is exposed alternately to thermal and mechanical loads. The number of load cycles ranges from 100 to 300.000, while the magnitude of the forces varies between 70 and 350 N [45].

Repeated casting of a dental alloy in the production of fixed dental restorations and metal skeletons of partial dentures leads to accelerated ageing of the dental alloy. Changes caused by combustion of microconstituents (deoxidants) and binding of some elements (if poured on an open flame, or in an induction apparatus in air), affect the microstructure of the dental alloy, and are responsible for changes in the mechanical and other characteristics of the alloy [46]. The altered physical and mechanical characteristics of the alloy affect the environment in which dental restorations are placed adversely, which, in turn, leads to adverse reactions that may affect the tissue response in the immediate environment.

Although opinions in the professional literature are divided on whether it is good to melt and cast one dental alloy more than once, we believe that obvious changes in the mechanical characteristics of the dental alloy and increasing patient sensitisation to metals suggest that manufacturers' instructions on melting and casting conditions must be followed. They generally say that fixed dental restorations should be made from dental alloys in their original form.

In accordance with ISO Standard 9693-2: 2016 (E), Sect. 6.4, a thermocyclic test (TC test) was carried out, where it is necessary to test samples prepared from dental alloys for cracks between or after the tests. The Standard specifies that the samples are exposed to boiling water, which is counted as 1. Thermal shock and cold water, which is counted to be 2. The TC test anticipates 10 repetitions of exposure to boiling and cold water. As a result, cracks in the samples will be indicated during, after, and 48 h after the TC test. All the tested materials were manufactured in accordance with ISO Standard 9693-2: 2016 (E), and with instructions for sample preparation according to Sect. 6.4. The DGAs presented in Table 2.7 were used in the TC test.

Test preparation. The making of dental constructions (samples) took place as follows: All samples were prepared identically, and in accordance with the ISO Standard 9693-2: 2016 (E). A cut model, ready for bridge modelling, waxed construction in the wax ready for inserting, waxed construction on the base of the cuvette, 4— Embedded construction in the cuvette. When the mass was hardened (30 min), the cuvette was placed in the furnace, preheated (400 °C–15 min), then the temperature was raised to 800 °C for 30 min. After 30 min at 800 °C the cuvette was ready for casting. After sand blasting, the alloy thickness on the facetted surfaceof the cast sample was 0.3–0.4 mm plus beads-mechanical retention- 0.4 mm (grinding to 0.2 mm). Casting temperatures for DGAs used in the preparation of appropriate dental constructions in accordance with the ISO Standard 9693-2: 2016 (E) were in the range 1000–1200 °C.

The application of an opaquer to the buccal surface of the bridge (sample) was performed on the already treated dental structure. After a few minutes the base opaque was applied with a Foundation brush and inserted for 1 min into the GC labolight box, which has three 27 W bulbs, where the coating became polymerised. This was followed by the application of a coloured opaquer with a brush, 2 × on the base opaque, and the sample was placed in the GC labolight box for 1 min each time. In the final phase, a GC gradia faceting material (dentin and enamel) was applied, and fixation was performed with a step light. After application, the coating was applied with Optiglaze liquid, after which all samples were placed in the GC labolight box for another 5 min. After polymerisation, the samples were washed under steam and sandblasted and polished, and coated with Optiglaze liquid, so as to ensure the gloss of the test samples in the gold dental alloy portion.

The equipment required for the thermocyclic test consisted of: (i) A container with cold water maintained at 0–20 °C by an external temperature control element; (ii) A container with boiling water maintained by an external heating element; (iii) A wire basket suitable for transferring the test specimen(s) rapidly between the containers, and preventing them from coming into direct contact with the walls or bottom while ensuring that the test piece(s) remain(s) submerged.

Procedure: The test objects were placed in the wire basket so that they were not in contact or under mechanical stress. The basket was placed in the boiling water; this process is counted as the first thermal shock [only when executing point b) for the first time]. Dwell time: (30 ± 5) s. The basket was transferred from the boiling water to the cold water (0–20 °C) within 3 s; this process is counted as the second thermal shock [only when executing point c) for the first time]. Dwell time: (30 ± 5) s. The basketis then returned to the boiling water [transfer time: < 3 s; dwell time: (30 ± 5) s]. This was repeated until the test objects had been quenched 10 times, or until (obvious) failure. After the last quenching the objects were and dried.

Test A: During the performance of the TC test, the samples are examined for cracks.

Test B: After the TC test, the samples are examined for cracks.

Test C: The examination is repeated 48 h after testing, to test for retarded cracking (if cracking was not observed immediately after the test).

The results of TC testing for the A, B and C tests for all samples (DGAs from Table 2.7.) were NO. In other words, no cracks appeared during all the measurements of the selected samples.

Characterisation techniques. The thickness of the tooth structure and the thickness of the tooth structure with the coating were measured using the Fino Dial Caliper (FINO GmbH) instrument at the centre of each tooth in the bridge.

The measurements of the coating thickness on the dental construction at the selected areas were carried out on an Electron Microscope, Quanta 200 3D (located in the University of Maribor, Faculty of Mechanical Engineering). The Quanta 200 3D is an environmental electron raster microscope with a tungsten cathode as the source of electrons; it allows work at various pressures and 100% humidity. By adjusting the pressure in the chamber, we obtain the conditions for the observation of different patterns. The Electron Microscope Quanta 200 3D has the potential for three basic modes of operation: High vacuum, low vacuum and environmental SEM mode. All measurements of the coating thickness on the dental construction were performed under the same conditions: Mode—high vacuum, Accelerating voltage—20 kV, Magnification—200 × , Detector—ETD, Sample tilt—20°. The measurements were made on areas where the coating thickness could be measured at the root of the bridge. On each tooth, it was attempted to measure at least two areas by 6 measurements, but, in some cases, this was not possible because of the inappropriate application of the alloy coating. The measurement areas are marked in the pictures. The coating thickness was measured before, immediately after, and 48 h after the TC test.

Based on the measured values, we can conclude that the average thickness ratio is: Cast tooth construction: Coating = 1: 3,5. The example of SEM examination and measurements on the dental structure for TC testing in the case of the Dentor S dental alloy is shown in Fig. 2.5. There are no visible defects in the tested samples of Dentor S, according to the TC test (tests A, B and C), the differences in the coating thicknesses are minimal, and can be attributed to operator error. It can be concluded that the tested samples of Dentor S fulfil the requirements of ISO Standard 9693-2: 2016 (E), in Sect. 6.4. The results of coating thickness measurement before, immediately after, and 48 h after the TC test for the Au dental alloy Dentor S are shown in Table 2.11.

Based on the extensive and detailed observations of samples or their selected areas with SEM microscopy, as well as measurements of the thickness of the required

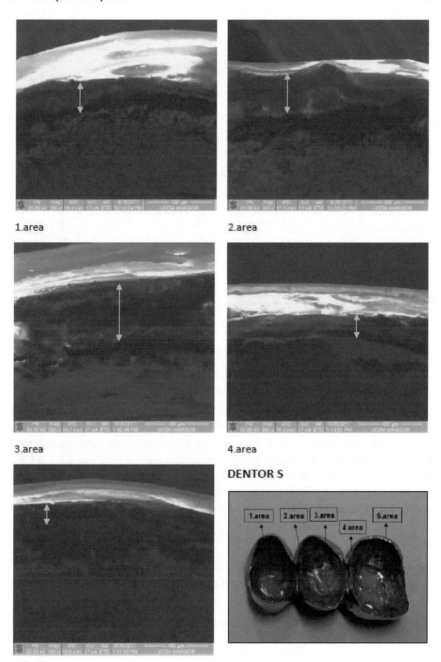

1.area

2.area

3.area

4.area

DENTOR S

5.area

Fig. 2.5 SEM micrographs of the areas in which the layer measurement was performed in the case of the Dentor S as Au dental alloy

Table 2.11 Thickness measurement of the coating before, immediately after, and 48 h after the TC test for Dentor S

1. Before the TC test [μm]

	1.area	2.area	3.area	4.area	5.area
Average	185.33	253.33	348.83	143.17	174.33
Min	165	220	331	125	161.00
Max	235	283	373	168	190
STDEV	26.77	20.16	17.75	14.47	9.37

2. After the TC test [μm]

	1.area	2.area	3.area	4.area	5.area
Average	181.50	240.83	380.83	181.17	173.33
Min	160	214	360	145	169
Max	223	279	413	219	178
STDEV	25.42	22.11	20.82	28.60	3.78

3. 48 h after the TC test [μm]

	1.area	2.area	3.area	4.area	5.area
Average	183.67	223.00	348.67	126.83	156.33
Min	170	195	314	96	145
Max	219	265	385	185	161
STDEV	18.63	25.23	28.46	35.26	6.77

coatings before and after TC, it can be concluded that all the tested dental alloys Au met the requirements of ISO 9693-2: 2016 (E) in Sect. 6.4. - Thermocyclic test. This is extremely important data for all manufacturers who have comparable dental alloys in their range, as well as for users such as dental laboratories and, last, but not least, for patients who come into direct contact with the investigated dental alloys in the long run (several years). Such tests ensure that dental alloys covered with appropriate coating are independent of temperature, and, in this way, suitable for clinical use.

2.2.3 Microstructure, Microhardness and Colour Comparison of Two Gold Dental Alloys

Materials and methods. Table 2.12 presents the composition of common yellow Au

Table 2.12 Composition of yellow and white gold dental alloys

Sample name	Chemical composition
Yellow Au	Au 57 wt. %, Ag 29 wt. %, Cu 14 wt. %
White Au	Au 56 wt. %, Ag 8 wt. %, Cu 23 wt. %, Zn 6 wt. %, Pd 7 wt. %

alloy and dental white Au alloy (Alloy Group 7, Indications B, D, L, see Table 2.2).

The metallographic samples were prepared in accordance with the appropriate Standards for metallographic preparation (brushing, polishing, etching). After cutting, the metallographic sample was polished on felts for polishing of varying fineness with a suitable polishing paste. After completion of the polishing and washing of the gold samples, the dried sample was etched with an etchant, namely 1 part: 1 g KCN + 10 ml H_2O and 2 parts: 1.1 $_g$ $(NH_4)_2S_2O_8$ + 10 ml H_2O. The etching time of the gold samples was 180 s. For all samples, a metallographic examination was performed in the longitudinal and transversal directions. The samples were observed with a light microscope—Nikon Epiphot 200 (Nikon, Japan) at different magnifications.

For identification of grain size, we used the Standard ASTM E112-12 Intercept method, which involves an actual count of the number of grains intercepted by a test line, or the number of grain boundary intersections with a test line, per unit length of test line, used to calculate the mean lineal intercept length, ℓ. ℓ is used to determine the ASTM grain size number, G. The precision of the method is a function of the number of intercepts or intersections counted. A precision of better than ±0.25 grain size units can be attained with a reasonable amount of effort. Results are free of bias; repeatability and reproducibility are less than ±0.5 grain size units.

On the samples, the HV1 microhardness was measured on a ZWICK 3212 device with 9804 N applied force and a 20 s load time. On each sample, the measurement was carried out in the longitudinal and transversal directions. In this way, minor or greater differences were found in the results.

A Spectrophotometer, Datacolor SF600 plus, was used for measuring the colour parameters of the Au samples with coatings. The instrument allows measuring of the remission spectrum of the incident light in 10 nm intervals within the visible part of the spectrum (360–700 nm).

Microstructure. Tables 2.13 and 2.14 present the optical microstructures and grain size of the selected studied gold dental alloys.

The metallographic examination of the yellow gold dental alloy microstructure has shown oxidation traces at the grain boundaries, a larger grain size compared to the white dental alloy microstructure, and the appearance of holes on the surface, which may be due to the preparation of the sample. The microstructure of the white gold dental alloy was the most homogeneous and finely grained. Measuring grain size is a complex process that depends on several factors, since the three-dimensional grain size is not constant, and the grain cross-section is random. Test methods include assessment procedures and rules for expressing the average grain size of a material, which consists essentially of only one phase. When comparing the grain size, the largest grain size was in a white dental gold alloy (10.05 μm).

Microhardness. We compared the microhardness of both gold dental alloys, as presented in Table 2.15. We saw that the white gold dental alloy had the highest HV1 microhardness value (297 in the transverse, or 295 in the longitudinal direction), compared with the sample from yellow gold dental alloy, and the yellow gold sample had the lowest value (235 in the transverse, or 225 in the longitudinal direction).

Table 2.13 Yellow gold dental alloy microstructure

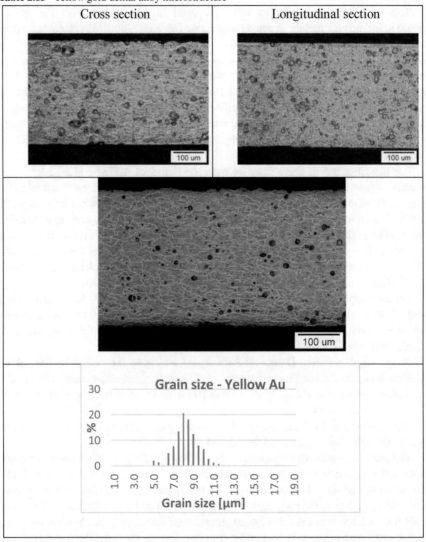

The reason for these different microhardness results is the fact that, in white gold dental alloy, the composition includes elements that increase the hardness (Zn, Pd). There is also an intermetallic mixture of copper and gold, which increases the strength, in contrast to the yellow gold, where there is a high percentage of gold, and hardness increases only with an intermetallic compound that forms from gold with copper.

Colour measurement. The measurement procedure includes calibration with black and white after the device is started, followed by each start-up when changing

Table 2.14 White gold dental alloy microstructure

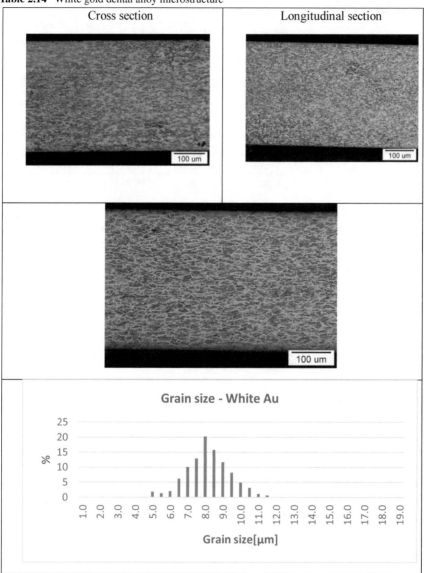

the measurement opening and every 8 h of measurements. This eliminates the effect of ageing the xenon flash on the measurement results. Each sample is placed and fixed with a special holder against the measuring aperture of the instrument. The sample is highlighted from the light source placed inside the instrument. The instrument measures the wavelengths of reflected and absorbed parts of the incident light. The measured data are stored into a computer database, and can be used by the special

Table 2.15 Microhardness results HV1

Yellow gold dental alloy		
	Cross-section	Longitudinal section
Mean value	235	225
White gold dental alloy		
	Cross-section	Longitudinal section
Mean value	297	295

Table 2.16 Colour measurement results

Sample	a*	b*	L*
Yellow	2,27,392	27,69,907	86,17,391
White	1,42,613	10,79,617	83,42,687

computer software for sorting and calculating the values of the colour parameters necessary for defining colour, whiteness, or computer colour matching.

Table 2.16 presents the colour measurement results of the white dental and yellow gold alloys.

From the colour measurement it can be seen that both gold dental alloys belong to pale yellow (very low value a *, and different value b *).

2.2.4   Corrosion Resistance of DGAs

(Electrochemical testing: The influence of chemical composition on the corrosion resistance) [47]

Corrosion resistance, besides biocompatibility, is an exceptionally important property of dental alloys. Numerous studies have used electrochemical techniques to evaluate the corrosion resistance of DGAs. Unfortunately, no reliable method has been proposed to classify DGAs based on their corrosion behaviour. The only existing Standard for the corrosion investigations of dental metallic materials is ISO 10271, which consists of three parts:

(1) A static immersion test, in a solution containing 10 g dm^{-3} lactic acid, 5.85 G dm^{-3} Sodium Chloride at pH $= 2.3$,

(2) An electrochemical test in 0.9 g dm^{-3} sodium chloride at pH $= 7.2$ in an inert atmosphere, and.

(3) A tarnish test in $3.1 \text{ g dm}^{-3} \text{ Na}_2\text{S} \times 9\text{H}_2\text{O}$. Nevertheless, there is no quantitative and qualitative description on how to interpret the results obtained by ISO 10271.

In this expertise, platinum and nine commercial DGAs with the contents of noble metals ranging from 27 wt. % to 97.6 wt. % were investigated according to the Standard ISO 10271:2009. Electrochemical testing was conducted in an argon purged solution of 0.9% NaCl at pH $= 7.2$. A three-compartment glass electrochemical cell with the volume 100 cm^3 was used, with an auxiliary platinum wire electrode and saturated calomel (SCE) as a reference electrode. Details are available in [47].

The obtained values, treating only Au, Pt, and Pd as noble, were compared with the American Dental Association (ADA) classification system (see Table 2.1). In that manner, the considered DGAs were classified into three existing groups, but with additional information related to electrochemical data.

As indicated, nine DGAs were tested, kindly supplied by Zlatarna Celje d.o.o., Slovenia, with the chemical composition given in Table 2.17. On the other hand, pure Pt was used as a well known corrosion-resistant model system. BIOKER is used mostly as a DGA for metal-ceramics, while other DGAs are used for faceting with artificial materials, as listed in Table 2.7.

From the obtained results it can be concluded that only the values of the breakdown potential and the current density at the potential of $E_{z,1} + 0.3$ V show some dependence in the tested DGAs. These values, summarised in Table 2.18., were analysed from the point of view of the total content of precious metals, $\Sigma N.M = Au + Pt + Pd$.

Finally, the values for breakdown potential and current density, $I_{0.3}$, at the potential $E_{0.3} = E_{z,1} + 0.3$ V were compared with the American Dental Association (ADA) compositional classification system, Table 2.19.

The results of this expertise allow division into three groups of stability:

Table 2.17 Chemical composition, density, Vickers hardness HV5 and the common applications of the tested DGAs [47]

Trade mark®	Chemical composition, wt.%	Noble metals,wt.%	Density, g cm^{-3}	HV5	Indication*
Bioker	Au 85.9; Pt 11.7; Zn 1.5; Ir, In < 1	97.6	18.9	210	3, 4, 5, 8
Aurodent 20	Au 83.3; Ag 10.0; Cu 5.9, Zn < 1	83.3	16.8	133	2, 3, 4, 8
Aurodent BIO	Au 83.3; Ag 10.0; Cu 6.7	83.3	16.8	145	1, 3, 4, 8
Dentor BIO	Au 78.3; Pt 5.0; Ag 10.0; Cu 6.7	83.3	16.2	134	3, 4, 6, 8
Dentor S	Au 75.5; Pd 1.2; Pt 4.4; Ag 11.0; Cu 6.7; Zn 1.2	81.1	16.4	232	3, 4, 5, 6, 8
Midor S	Au 46.0; Pd 6.0; Ag 39.5; Cu 7.5; Zn, Ir < 1	52.6	12.9	265	3, 4, 5, 6, 8
Midor SE	Au 40.0; Pd 4.0; Ag 47.0; Cu 7.5; Zn, Ir, < 1	44	12.4	248	3, 4, 8
Auropal S	Au 10.5; Pd 21.0; Ag 58.2; Cu 9.3; Zn < 1	31.5	11.2	267	3, 4, 5, 6, 8
Auropal SE	Au 2.0; Pd 25.0; Ag 64.0; Cu 8.0. Zn < 1	27	10.7	250	3, 4, 5, 6, 8

*Indication: 1—Inlay; 2—MOD inlay; 3—Crowns; 4—Small bridges; 5—Large bridges; 6—Milling, telescopes; 8—Pins

Table 2.18 The dependence of composition and total content of noble metals $\Sigma N.M = Au + Pt + Pd$ on the current density at potential $E_{z,1} + 0.3$ V, and breakdown potential, E_p [47]

Dental Alloy	Chemical composition	$\Sigma N.M$	$j_{0.3}$, $\mu A\ cm^{-2}$	E_p, V
Pt	Pt 100	100	0.28	1.02
Bioker	Au 85.9; Pt 11.7; Zn 1.5; Ir. In < 1	97.6	0.23	0.87
Aurodent 20	Au 83.3; Ag 10.0; Cu 5.9	83.3	0.8	0.78
AurodentBIO	Au 83.3; Ag 10.0; Cu 6.7	83.3	1.1	0.83
Dentor BIO	Au 78.3; Pt 5.0; Ag 10.0; Cu 6.7	83.3	1.4	0.81
Dentor S	Au 75.5; Pd 1.2; Pt 4.4; Ag 11.0; Cu 6.7; Zn 1.2	81.1	0.52	0.82
Midor S	Au 46.0; Pd 6.0; Ag 39.5; Cu 7.5; Zn, Ir < 1	52.6	2.3	0.55
Midor SE	Au 40.0; Pd 4.0; Ag 47.0; Cu 7.5; Zn, Ir < 1	44	3	0.44
Auropal S	Au 10.5; Pd 21.0; Ag 58.2; Cu 9.3; Zn < 1	31.5	3.85	0.31
Auropal SE	Au 2.0; Pd 25.0; Ag 64.0; Cu 8.0; Zn < 1	27	4.9	0.32

Table 2.19 American Dental Association (ADA) classification and the values of current density at the potential of $E_{z,1} + 0.3$ V (SCE), and breakdown potential, E_p, with a suggested group of stability [47]

ADA Classification	Noble metal contents, wt.%	$j_{Ez,1+0.3\ V(SCE)}$ $\mu A\ cm^{-2}$	E_p V versus SCE	Stability group
High-noble	$\Sigma N.M > 60$, Au > 40	< 2	> 0.6	I, most stable
Noble	$\Sigma N.M > 25$	2–5	0.3–0.6	II, stable
Predominantly base	$\Sigma N.M < 25$	> 5	< 0.3	III, unstable

(I) most stable Dental alloys, or the high-noble dental alloys with the chemical composition of at least 60 wt. % of noble metals, and with gold content at least 40 wt. %; which should have $I_{0.3} < 2$ $\mu A\ cm^{-2}$ and $E_p > 0.6$ V versus SCE.

(II) stable alloys Dental alloys, or the noble dental alloys with chemical composition with at least 25 wt. % to 60 wt. % of noble metals, which should have $I_{0.3}$ between 2 and 5 $\mu A\ cm^2$, and E_p between 0.6 V and 0.3 versus SCE.

(III) unstable alloys Dental alloys, or the predominantly base dental alloys, with chemical composition less than 25 wt. % noble metals, which should have $I_{0.3} > 5$ $\mu A\ cm^{-2}$ and $E_p < 0.3$ V versus SCE.

In addition to these results, based on the EDS, SEM and ICP-MS analyses, it was confirmed that all alloys possess high corrosion stability, at open circuit potentials and under anodic polarisations.

It should be known that procedures defined by ISO 10271 gave limited results. From the five measured corrosion parameters, only two showed some dependencies on the noble metal content. It is, therefore, necessary to establish a more rigorous procedure and classification of DGAs.

2.2.5 The Influence of Microstructures on the Corrosion and Biocompatibility in Vitro (An Example of High Noble Au-Pt Dental Alloys) [48]

The clinical experience has indicated that, from various types of dental alloys for porcelain fused to metal (PFM) restorations, Au-Pt based high noble dental alloys are extremely successful, with a particularly strong and highly reliable bond between the ceramic and the metal.

High Au concentrations are obliged to provide biocompatibility and rich Pt contents to raise the melting range sufficiently above the porcelain firing temperature. Thus, it prevents distortion during the porcelain's application. The addendum of Zn lowers the liquid alloy's surface tension, so the liquid metal could be cast into very thin sections. Zn as a deoxidant, together with Ir and In, protect other metals from oxidation. On the other hand, Ir and In as micro-alloying metals are added to establish a thin oxide film at the alloy's surface during the porcelain firing cycle. The Rh enhances both strength and colour.

In this presentation, the influence of the microstructures of two high noble Au-Pt dental alloys with similar composition on their corrosion and biocompatibility, were compared in vitro. We had an Au-Pt I dental alloy (86.9 wt.% Au, 10.4 wt.% Pt, 1.5 wt.% Zn and 0.5 wt.% Ir + Rh + In) and an Au-Pt II dental alloy (87.3 wt.% Au, 9.9 wt.% Pt, 1.7 wt.% Zn and 0.5 wt.% Ir + Rh + In). The slightly higher Zn content in the Au-Pt II dental alloy was chosen to improve the bonding strength between the porcelain and the alloy, and the mechanical properties for the tooth's metallic substitute.

After a preliminary examination, our viability study on L929 cells found that Au-Pt I and Au-Pt II dental alloys, despite being of similar composition, differed notably in their biocompatibility. We assumed that the differences in microstructure and distribution of microalloying elements within the phases, especially Zn, were responsible for this result. Therefore, we organised our examinations further as follows:

1. To evaluate metal ion release from two Au-Pt dental alloys into a culture media;

2. To examine the cytotoxic effect using a standard viability assay on L929 cells and more sensitive T-cell function and apoptosis assays;

3. To test whether the released Zn ions in the conditioning media would cause the same biological acitivity as the exogenous added Zn;

4. To contrast the microstructure of the two Au-Pt dental alloys before and after conditioning.

After the entire procedure, it was amazing that the Au-Pt II dental alloy, although possessing better mechanical properties than the Au-Pt I dental alloy, exerted higher adverse effects on the viability of L929 cells and the suppression of rat thymocyte functions, such as proliferation activity, production of Interleukin-2 (IL-2), expression of IL-2 receptors and activation—induced apoptosis after stimulation of the cells with Concanavalin-A. These results correlated with the higher release of Zn ions in the culture medium. As Zn^{2+}, at the concentrations which were detected in the alloy's culture media, showed a lesser cytotoxic effect than the Au-Pt conditioning media, we concluded that Zn was probably not the only element responsible for the alloy's cytotoxicity. Microstructural characterisation of the dental alloys performed by means of Scanning Electron Microscopy in addition to Energy Dispersive X—ray and X—ray diffraction analyses, showed that Au-Pt I is a two-phase dental alloy containing a dominant Au-rich α_1 phase and a minor Pt-rich α_2 phase. On the other hand, the Au-Pt II dental alloy additionally contained three minor phases: $AuZn_3$, Pt_3Zn and $Au_{1.4}Zn_{0.52}$. The highest content of Zn was identified in the Pt_3Zn phase. After conditioning the Pt_3Zn and $AuZn_3$ phases disappeared, suggesting that they were predominantly responsible for the Zn loss, lower corrosion stability and subsequent lower biocompatibility of the Au-Pt II alloy [48].

The microstructural analysis of the Au-Pt alloys was performed by SEM, EDX and XRD. The SEM and elemental mapping results are presented in Fig. 2.6. The Au-Pt I alloy, before the biocompatibility test, was composed of an Au—dominant α_1 phase (98.44 wt. %) and a minor Pt—dominant α_2 phase (1.56 wt. %). On the other hand, the Au-Pt II alloy contained additionally three minor phases ($AuZn_3$, Pt_3Zn and $Au_{1.4}Zn_{0.52}$), and the calculated mass ratio (wt.%) was $\alpha_1:\alpha_2:AuZn_3:Pt_3Zn:Au_{1.4}Zn_{0.52} = 95.55:2.72:0.98: 0.24:0.5$. The phases in these dental alloys were identified and analysed according to the latice parameters and their comparison with pure Au and Pt and $AuZn_3$, Pt_3Zn and $Au_{1.4}Zn_{0.52}$ compounds. The α_1 phase of the Au-Pt I was richer in Pt and poorer in Au compared to the α_1 phase of the Au-Pt II. The opposite results were obtained for the α_2 phase. The content of Zn in both the α_1 and α_2 phases of the Au-Pt II alloy was higher compared to the Au-Pt I. The other three minor phases were enriched in Zn, and the highest content was detected in the Pt_3Zn phase, followed by $Au_{1.4}Zn_{0.52}$ and $AuZn_3$ [48].

After conditioning, the α_2 phase was visible in the microstructure of the Au-Pt I dental alloy, whereas, in the Au-Pt II microstructure, this phase was much bigger and more distinctive. The EDX-point analysis for all the detected phases showed that the α_1-phase regions became poor in Zn, and within them there was no Rh, In and Ir. The α_2 phases also contained smaller portions of Zn and Rh, and, consequently, higher concentrations of Au and Pt. The diffraction patterns for the Au-Pt I dental alloy (Fig. 2.6) showed the presence of another group of peaks (marked α_{2-s}), shifted from the α_2 phase by approximately 1°, most probably as a consequence of micro-segregation in the interior of the α_2 phase. The calculated mass ratio (wt. %) for the phases of Au-Pt I after conditioning was: $\alpha_1:\alpha_2:\alpha_{2-s} = 90.82:4.50:4.67$.

After conditioning, the $AuZn_3$ and Pt_3Zn phases disappeared from the surface of the Au-Pt II alloy, such that the calculated mass ratio for the existing phases was $\alpha_1:\alpha_2: Au_{1.4}Zn_{0.52} = 94.28:4.67:1.05$.

In the end, it can be concluded that: (a) The microstructure, but not the composition of a high noble Au-Pt dental alloy, is connected with its corrosion properties and biocompatibility in vitro; (b) The presence of the $AuZn_3$ and Pt_3Zn phases in the dental alloy led to lower corrosion stability; (c) T-cell functional tests are more sensitive for evaluating the adverse effect of Au-Pt dental alloys than a conventional MTT assay on L929 cells; (d) The influence of whole Au-Pt dental alloy extracts on the biocompatibility is more complex than the effect of Zn alone, although Zn

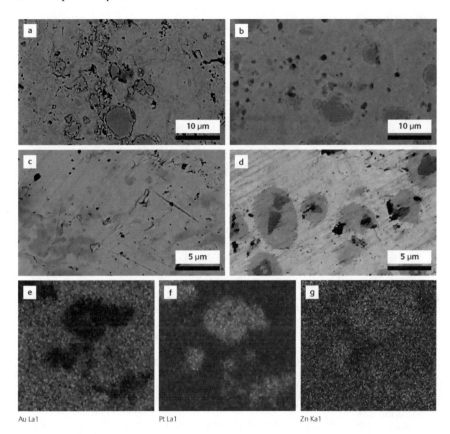

Fig. 2.6 The SEM microstructure of the Au-Pt dental alloys before and after conditioning, and EDX chemical mapping of the elements in the region of the Au-Zn and Pt-Zn phases in the Au-Pt II dental alloy. **a** SEM of Au-Pt I before conditioning; **b** SEM of Au-Pt II before conditioning; **c** SEM of Au-Pt I after conditioning; **d** SEM of Au-Pt II after conditioning; **e** EDX mapping of Au; f) EDX mapping of Pt; g) EDX mapping of Zn [48]

was the only detectable element released from the dental alloys. Generally, these results contradict those published by other authors, [49, 50] who showed that dental alloy extracts had lesser cytotoxic effect than the same concentrations of Zn salt. The difference could be explained by the different compositions and microstructures of alloys, different modes of alloy conditioning, different cell culture media and different biological parameters that were monitored.

Finally, the microstructural and XRD analyses of dental alloys before and after conditioning, in combination with the analysis of element release from the alloys, could be a new approach in explaining the results of biocompatibility assays.

References

1. Becker MJ, Turfa JMI (2017) The Etruscans and the History of Dentistry/The Golden Smile through the Ages. Taylor & Francis
2. Knosp H, Holliday RJ, Corti CW (2003) Gold in dentistry: alloys, uses and performance. Gold Bull 36(3):93–102
3. Phillips RW (1973) Science of dental materials, 7th edn. WB Saunders, Philadelphia
4. Craig RG, Powers JM (2002) Restorative dental materials, 11th edn. Mosby, St. Louis
5. Yasuda K, Hisatsune K (1993) Microstructure and phase transformations in dental gold alloys/determination of a coherent phase diagram. Gold Bull 26(2):50–66
6. German RM (1980) Gold alloys for porcelain-fused-to-metal dental restorations. Gold Bull 13(2):57–62
7. Dental Alloy with a High Gold Content, European Patent EP 0691123 (1997)
8. Rosenstiel S, Land M (Eds) (2015) Contemporary fixed prosthodontics, 5th edn. Mosby
9. Wataha JC (2002) Alloys for prosthodontic restorations. J Prosthet Dent 87(4):351–363
10. No authors listed (1984) Classification system for cast alloys, council on dental materials, instruments, and equipment. J Am Dent Assoc 109(5):766
11. Wataha JC, Messer RL (2004) Casting alloys. Dent Clin N Am 48(2):499–512
12. Upadhyay D, Panchal MA, Dubey RS, Srivastava VK (2006) Corrosion of alloys used in dentistry: a review. Mat Sci and Eng A 432:1–11
13. Nierlich J, Papageorgiou SN, Bourauel C, Hültenschmidt R, Bayer S, Stark H, Keilig L (2016) Corrosion behavior of dental alloys used for retention elements in prosthodontics. Eur J Oral Sci 124(3):287–294
14. German RM, Guzowski M, Wright DC (1980) The colour of Au-Ag-Cu alloys: quantitative mapping of the ternary diagram. Gold Bull 13:113–116
15. Cretu C, van der Lingen E (1999) Coloured gold alloys. Gold Bull 32:115–126
16. Williams DF (2008) On the mechanisms of biocompatibility. Biomaterials 29(20):2941–2953
17. Wintermantel E, Ha SW (1998) Biokompatible Werkstoffe. Springer Verlag
18. Wataha JC (2000) Biocompatibility of dental casting alloys: a review. J Pros-thet Dent 83(2):223–234
19. Wataha JC, Nelson SK, Lockwood PE (2001) Elemental release from dental casting alloys into biological media with and without protein. Dent Mater 17(5):409–414
20. Al-Hiyasat AS, Bashabsheh OM, Darmani H (2002) Elements released from dental casting alloys and their cytotoxic effects. Int J Prosthodont 15(5):473–478
21. Anusavice KJ (1996) Philips' science of dental materials. Saunders, Philadelphia, ch. 15(20):23
22. Grieve AR, Saunders WP, Alani AH (1993) The effects of dentine bonding agents on marginal leakage of composite restorations, long-term studies. J Oral Rehabil 20(1):11–18
23. Stamenković D, Čairović A, Rudolf R, Radović K, Đorđević I (2009) The effect of repeated casting on the biocompatibility of a dental gold alloy. Acta veter-inaria 59(5/6):641–652
24. Leinfelder KF, O'Brien WJ, Taylor DF (1972) Hardening of dental gold-copper alloys. J Dent Res 51(4):900–905
25. Yasuda K, van Tendeloo G, van Landuyt J, Amelinckx S (1986) High-resolution electron microscopic study of age hardening in a commercial dental gold alloy. J Dent Res 65(9):1179–1185
26. Cullity BD (1978) Elements of X Ray diffraction, reading, Addison-Wesley, chap 13, MA
27. Reisbick MH, Brantley WA (1995) Mechanical property and microstructural variations for recast low-gold alloy. Int J Prosthodont 8(4):346–350
28. Winn H, Udoh K, Tanaka Y, Hernandez RI, Takuma Y, Hisatsune K (1999) Phase transformations and age-hardening behaviours related to Au3Cu in Au-Cu-Pd alloys. Dent Mater J 18(3):218–234
29. Watanabe I, Atsuta M, Yasuda K, Hisatsune K (1994) Dimensional changes related to ordering in an AuCu-3 wt.% Ga alloy at intraoral temperature. Dent Mater 10(6):369–374
30. Kim H-II, Seol HJ, Bae DH (1999) Isothermal age-hardening behaviour in a Au-1.6 wt% Ti Alloy. Dent Mater J 18(1):32–41

31. Brick RM, Pense AW, Gordon RB (1977) Structure and properties of engineering materials. McGraw-Hill, New York, chap 2:10
32. Brantley WA, Cai Z, Carr AB, Mitchell JC (1993) Metallurgical structures of as-cast and heat-treated high-palladium dental alloys. Cells Mater 3(1):103–114
33. Vermilyea SG, Huget EF, Vilca JM (1980) Observations on gold-palladium-silver and gold-palladium alloys. J Prosthet Dent 44(3):294–299
34. Baran GR (1983) The metallurgy of Ni-Cr alloys for fixed prosthodontics. J Prosthet Dent 50(5):639–650
35. Morris HF, Asgar K, Rowe AP, Nasjleti CE (1979) The influence of heat treatments on several types of base-metal removable partial denture alloys. J Prosthet Dent 41(4):388–395
36. Bridgeport DA, Brantley WA, Herman PF (1993) Cobalt-chromium and nick-el-chromium alloys for removable prosthodontics, part 1: mechanical properties of as-cast alloys. J Prosthodont 2(3):144–150
37. Yasuda K, Ohta M (1982) Difference in age-hardening mechanism in dental gold alloys. J Dent Res 61(3):473–479
38. Bayne CS (2005) Dental biomaterials: where are we and where are we going? J Dent Educ 69(5):571–585
39. Čolić M, Rudolf R, Stamenković D, Anžel I, Vučević D, Jenko M et al (2010) Relationship between microstructure cytotoxicity and corrosion properties of a Au-Al-Ni shape memory alloy. Acta Biomater 6(1):308–317
40. Grill V, Sandrucci MA, Di Lenada R, Canderano M, Narducci P, Bareggi R, Martelli AM (2000) Biocompatibility evaluation of dental metal alloys in vitro: expression of extracellular matrix molecules and its relationship to cell proliferation rates. J Biomed Mat Res 52(3):479–487
41. van Noort R (1994) Introduction to dental materials. Mosby, London
42. Wang JN, Liu WB (2006) A Pd-free high gold dental alloy for porcelain bonding. Gold Bull 39(3):114–120
43. Seol HJ, Kim GC, Son KH, Kwon YH, Kim H-Il (2005) Hardening mechanism of an Ag–Pd–Cu–Au dental casting alloy. J Alloys Compd 387(1–2):139–146
44. Kournetas N (2005) Impact of artificial ageing process on the wear resistance of dental materials. Dissertation, University of Tübingen
45. Morresi AL, D'Amario M, Capogreco M, Gatto R, Marzo G, D'Arcangelo C, Monaco A (2014) Thermal cycling for restorative materials: does a standardized protocol exist in laboratory testing? A literature review. J Mech Behav Biomed Mater 29:295–308
46. Vasiliu RD, Porojan SD, Bîrdeanu MI, Porojan L (2020) Effect of thermocycling, surface treatments and microstructure on the optical properties and roughness of CAD-CAM and heat-pressed glass ceramics. Materials (Basel) 13(2):381
47. Grgur BN, Lazić V, Stojić D, Rudolf R (2021) Electrochemical testing of noble metal dental alloys: the influence of their chemical composition on the corrosion resistance. Corrosion Sci 184:109412
48. Čolić M, Stamenković D, Anžel I, Lojen G, Rudolf R (2009) The influence of the microstructure of high noble gold-platinum dental alloys on their corrosion and biocompatibility in vitro. Gold Bull 42(1):34–47
49. Schmalz G, Langer H, Schweikl H (1998) Cytotoxicity of dental alloy extracts and corresponding metal salt solutions. J Dent Res 77(10):1772–1778
50. Waters M (2003) Dental Materials in Vivo, Ageing and Related Phenomena. Br Dent J 195:722. https://doi.org/10.1038/sj.bdj.4810850

Chapter 3
Gold Nanoparticles

3.1 Gold Nanoparticles' Characteristics and Possibilities of Use

Materials with one dimension below 100 nm, or nanomaterials [1], have different properties when compared to their bulk material counterparts. These properties come from their large surface to volume ratio, altering their physical and chemical properties. Their increased surface activity makes them useful in various areas, from Electronics, Chemistry, Biotechnology and Medicine, to be used as electrochemical devices, for sensor applications and catalysis, energy conversion and for storage [1–4].

3.1.1 Optical Properties and Surface Plasmon Resonance

Metallic nanoparticles, such as gold, silver and copper nanoparticles, are especially useful due to a surface effect called Surface Plasmon Resonance (SPR) [2, 5, 6]. This effect causes the collective coherent oscillations of free electrons in the conduction band of a metal when excited by an interactive electromagnetic field at a metal/dielectric interface. In other words, when a metal nanoparticle in a dielectric medium is subjected to light, the oscillations of conduction electrons around the surface of the particle causes differences of the charges in the atomic structure, and creates a dipole oscillation in the direction of the electrical field of incident light [7] (Fig. 3.1). These created charge density oscillations are called surface plasmon polaritons [6]. Four factors determine the frequency of oscillations: The density of the electrons, the effective electron mass, and the shape and size of the charge distribution [8].

Localised Surface Plasmon Resonance—LSPR (localised to a small area, such as a nanoparticle) enhances the optical properties of nanoparticles, such as absorption and

© The Author(s), under exclusive license to Springer Nature Switzerland AG 2022   53
R. Rudolf et al., *Dental Gold Alloys and Gold Nanoparticles for Biomedical Applications*,
SpringerBriefs in Materials,
https://doi.org/10.1007/978-3-030-98746-6_3

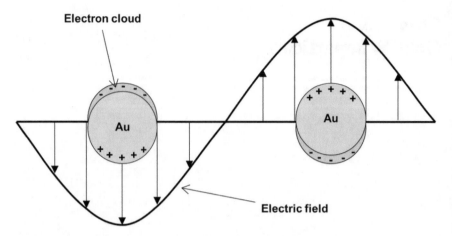

Fig. 3.1 Representation of GNPs with oscillating electrons, adapted from [7, 8]

scattering of incident light [7]. Strongly absorbed light is also transformed rapidly into heat. The absorption of light by nanoparticles can be measured with a UV–Vis spectrometer. The wavelength and intensity of the resonance band depend on the density of electron charge on the nanoparticle's surface. The electron charge density is, in turn, dependent upon the characteristic properties of the nanoparticle: Metal type, size, shape, structure, composition, and the dielectric properties of the surrounding medium. Theoretically, this was described as early as 1908 by Mie [9].

SPR causes increased absorption and scattering of visible light, making suspensions of GNPs with spherical shapes and sizes below 100 nm a red colour [1]. The SPR wavelength of spherical GNPs is around 520 nm, absorbing light in the green area of the visible light spectrum and scattering red light [7, 10]. Their colour changes with the different shapes, sizes and agglomeration, which alters the SPR effect, and, thus, the absorption and scattering of different wavelengths of visible light (Fig. 3.2). Silver nanoparticles of similar dimensions and shapes have a yellow appearance, with an SPR wavelength of about 400 nm [11], absorbing blue light and scattering yellow light.

With noble metal nanoparticles, particularly GNPs, the SPR effect is more enhanced as compared to other metal nanoparticles. Noble metals are also chemically inert, oxidation-free, and show high biocompatibility, which is critical for biomedical applications [13]. The different shapes of GNPs, such as nanorods, may also be tuned to other SPR wavelengths by changing their aspect ratio (length divided by width) [14]. The unique photoelectric properties of GNPs have been studied and applied in fields such as Sensing probes, Therapeutic agents and Drug delivery systems, Contrast enhancement of X-ray Computed Tomography, Diagnostics, Plasmonic bio-sensing, Colorimetric sensing, Tissue engineering, Photo-induced therapy, and Cancer therapy in biological and pharmaceutical applications [10, 15].

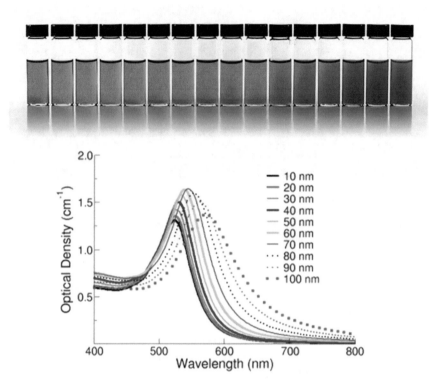

Fig. 3.2 Suspensions of GNPs with different colours, depending on their size. The shade of the red colour turns to purple, and then blue as the nanoparticle size increases. Below is a chart of the SPR extinction spectra (extinction = light that is not transmitted through a sample, or the sum of absorption and scattering of visible light) for GNPs of different sizes, adapted from nanoComposix [12]

3.1.2   Sizes and Shapes of GNPs

GNPs and larger gold particles are commercially available, or are included in products with sizes ranging from 1 nm to 8 μm, in shapes such as nanospheres, nanorods, nanocubes, nanobranches, nanoprisms, nanobipyramids, nanoflowers, nanoshells, nanowires, nanocages and irregular shapes [16, 17] (Fig. 3.3). The various sizes and shapes influence the surface to volume ratio, and, thus, the SPR and chemical activity of GNPs, changing the practicality of their use for different applications. For example, a high surface to volume ratio is needed for catalysis [18, 19], while, for cosmetics, other properties are more important, such as their ability to interact with the skin barrier, enhancing delivery and improving the skin permeability of the high-molecular-weight active agents present in skin creams [16]. Sizes are an important aspect and field of study for GNPs used in biomedical areas, where it has been shown

Fig. 3.3 TEM images of Au nanoparticles of different shapes and sizes: **a** nanospheres, **b** nanocubes, **c** nanobranches, **d–f** nanorods with different aspect ratios, **g–j** nanobipyramids with different aspect ratios. Reprinted (adapted) with permission from Chen et al., Shape- and size-dependent refractive index sensitivity of gold nanoparticles, Langmuir [22]. Copyright 2008 American Chemical Society

that these physical properties have a great impact on their functional properties in these areas, such as accumulation in different tissues [20, 21].

The sizes and morphologies of the GNPs are dependent on their production method. Anisotropic shapes are usually formed when a stabilising polymer binds to one crystal face and causes the GNP to grow faster in one direction. Several methods have been developed to control the shapes of GNPs, utilising the preferential growth of crystals in various directions by changing the product parameters and conditions [17, 23, 24]. Ultimately, the sizes and shapes of GNPs are determined by the application in which they are being used.

3.1.3 Stability, Nanoparticle Surfaces and Functionalisation

The high surface energy of bare GNPs promotes the joining of smaller nanoparticles into larger aggregates or agglomerates. The smaller particles have a higher tendency to aggregate, due to their increased surface to volume ratio and higher surface energy as compared to larger nanoparticles. The joining of two smaller, primary particles into larger aggregates or agglomerates is due to chemical or physical forces.

Primary particles joined closely to each other by chemical forces (covalent or ionic) form necks through coalescence, and are so-called hard agglomerates or aggregates, which are difficult to break apart into primary particles (Fig. 3.4). In contrast, the so-called soft agglomerates (simply called agglomerates) are primary particles or aggregates attached loosely by physical forces (electrostatic or van der Waals). These can be broken up into primary particles by applying force or energy, for example, by subjecting the particles to ultrasound [25].

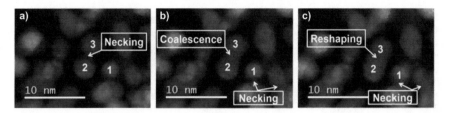

Fig. 3.4 The formation of necks between particles and coalescence of two GNPs during TEM observation shown from panel **a–c**, the coalescence and aggregation of the nanoparticles is promoted by the high energy electron beam, adapted from [26]

The beneficial properties of GNPs, such as SPR, change when aggregates or agglomerates are formed. In order to prevent the joining of GNPs, stabilisation of their surfaces is needed to prevent the loss of the SPR function. Stabilisation depends on the application of the GNPs, and is influenced by several factors, such as particle sizes, concentration, surfactants, and the local environment. GNPs may be stabilised electrostatically or sterically with ligands and surface shells (organic or inorganic) [13].

With charge stabilisation, the aggregation is prevented by the surface charge of the nanoparticles. The stability of charge stabilised particles can be measured with zeta potential, a parameter that provides information on the net charge of particles in a liquid environment [27]. Nanoparticles with a zeta potential greater than 20 mV or lower than −20 mV are considered to be stabile through electrostatic repulsion of the particles in the solution. If the particle surface has a positive charge then negative ions are attracted to the surface, and if there is a negative charge on the particle surface positive ions are attracted to it. The slipping plane is at the distance from the particle surface where the ions move with the particle. Zeta potential is measured at the slipping plane. Consequently, the zeta potential is highly dependent on local conditions, such as pH values and the presence of other molecules or contaminants in the solution, such as salts (Fig. 3.5).

Sterically stabilised GNPs have surfactant molecules bound to the nanoparticle surface by physisorption or by chemical bonds (polymers, ligands, proteins, surface shells) [13]. The surfactant molecules present a double layer of charge, providing steric hindrance and preventing coalescence and agglomeration of particles. The surfaces of GNPs are influenced greatly by the surrounding environment. Environments rich with salts or a complex biological matrix affects the steric double layer and will break it, causing GNP aggregation. For this reason, there are several methods for providing stabilisation with several surfactant molecules, and the choice of surfactants depends mainly on the application where the nanoparticles are being used, with retaining their plasmonic properties in the medium of application. The most common are Sodium Citrate (also used as a reducing agent in many synthesis methods), Polyvinylpyrrolidone (PVP) and thiol-functionalised groups, such as Polyethylene Glycol (PEG). GNPs without an appropriate stabiliser cannot maintain their structure, and will coalesce and aggregate or dissolve, losing their plasmonic effects [28].

Fig. 3.5 Representation of zeta potential of a particle with negative surface charge in a colloid suspension. Positive ions are populating the Stern layer, the slipping plane represents the distance from the surface where the ions move with the particle and zeta potential is measured

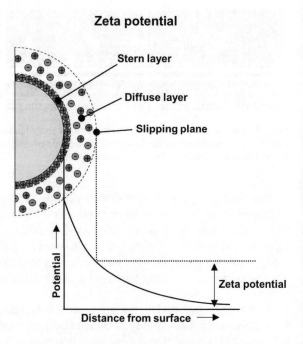

In order to provide a functional property for the GNPs, the initial stabilisation surfactant may be replaced by other stabilisers specific to an application's needs, while retaining the plasmonic properties and performing the desired function. This modification of the GNP surface is called functionalisation. The GNPs may be functionalised with a number of functional substances, such as molecules with thiols, amines and polymers. GNP surfaces can also associate directly with molecules containing proteins, drugs, antibodies, enzymes, nucleic acids (DNA or RNA) and fluorescent dyes, for various medical applications with biological activities and functions [29]. Additionally, several stabilising and functionalisation steps are used in many studies in order to maximise the nanoparticles' plasmonic abilities and perform other critical functions in the given application [13] (Fig. 3.6).

In contrast to ligand molecules that are bound to the GNP surfaces chemically or physically, shells are chemically and physically robust structures covering the nanoparticle's core homogeneously [13]. The most common shell types are silica, organic polymer and metallic. Shells may not be replaced as easily by other substances, as is the case with surface ligands, and can provide stabilisation for the cores through steric hindrance or electrostatic repulsion (silica and polymers), or by chemical interactions (metal shells). They may even provide additional functionality to the GNPs, such as magnetic properties or enhanced catalytic activity. A shell may be formed by the synthesis process or by replacing a ligand surfactant [30].

The main challenge with stabilisation and functionalisation is preserving the plasmonic properties of GNPs in their local environment of use, with adding the required functional capabilities. The stabilisation approach is, thus, dependent on the synthesis

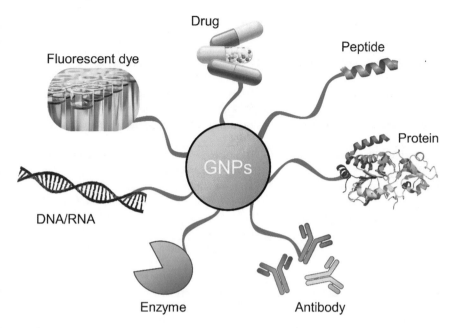

Fig. 3.6   Various molecules on the surface of GNPs, adapted from [17]

method and subsequent use. The initial stabilising surfactants and ligands that are loosely bound are also important for characterisation of GNPs, the determination of their morphologies and other properties. The subsequent ligand exchange with more tightly bound surfactants is then utilised for use of nanoparticles in their intended environment, for enhancing their stability and preserving, or even improving, their plasmonic functionality.

3.1.4   Characterisation of GNPs

To obtain a complete description of the GNP properties, several different measurement and analysis techniques are employed, depending on the area of use. Ordinarily, more than one technique is needed for determining the practicability and efficiency of GNPs in their respective use cases (Table 3.1).

Table 3.1 Characterisation methods for GNPs, adapted from [31]

Analysis method	Measurement consideration
Dynamic Laser Light Scattering (DLS)/Particle size analyser	Measurement of the particle size and size distribution of nanoparticles in liquid solutions or suspensions
Zeta potential analyser	Measurement of the surface charge of nanoparticles in aqueous solutions or suspensions
Scanning Electron Microscope (SEM)	Examination of a nanoparticle's surface, size and shape
Transmission Electron Microscope (TEM)	Determination of the surface property, size and shape morphology of nanoparticles
Energy Dispersive X-ray analysis (EDX)	Semi-quantitative analysis of the chemical composition, available in SEM and TEM instruments
X-ray Diffraction (XRD)/Electron diffraction	Determination of the crystallographic structure of nanoparticles
Atomic Force Microscope (AFM)	Measurement of the shape and surface morphology (including friction and softness) of nanoparticles with high lateral and vertical resolutions
Laser Scanning Confocal Microscope (LSCM)	Non-invasive measurement of nanoparticles' morphology in 3D, investigating the migration of nanoparticles into a bio-barrier
Surface area analyser and pore size analyser	Determination of single and multipoint surface area analysis, multigas capability and full adsorption capability for nanoparticles
X-Ray Photoelectron Spectroscope (XPS, ESCA)	Providing important chemical composition (both elemental and chemical states) information on nanoparticles' surfaces
Fourier Transform Infrared Spectroscope (FTIR)	Assisted analytical tool for chemical composition of nanoparticles' surfaces
Differential Scanning Calorimetry (DSC)	Providing thermal analysis (and component interactions) of nanoparticles and related materials during the fabrication process
High Performance Liquid Chromatography (HPLC)	Detection, separation and quantification of nanoparticles/nanomaterials with different particle sizes
Ultraviolet–visible spectroscopy (UV–Vis)	Measurement of the absorption or reflectance of nanoparticles in the UV and visible spectrum of light

3.2   GNP Current and Potential Applications

The properties of GNPs make them beneficial in a number of areas, while the widest research interest for their use is in the Biomedical field. The following sections describe briefly some of the application uses of GNPs in catalysis, electronics, sensors, cosmetics and food. A subsequent chapter is dedicated to a more detailed insight into the uses of GNPs in biomedical and other health related applications.

3.2.1 Catalysis

As GNPs are being researched and used more extensively, their catalytic properties are becoming more noteworthy, also with exploiting the plasmonic effects with catalysis through plasmon-mediated photocatalysis [32]. GNPs have the ability to lower the activation energy of some reactions, and increase the rate of these reactions and the yield of the desired products [14]. In contrast to inherently inert bulk gold and other noble metals, small particles of these metals have been found to be very catalytically active, and research in this area is focused on the influence of nanoparticles' morphology, shapes, edges and surface areas on their catalytic activity [33]. It was shown that nanoparticles that have shapes with more corners and edge atoms have a higher catalytic reactivity than nanoparticles with fewer corners and edge atoms. Additionally, the catalytic activity is being investigated for homogeneous and heterogeneous systems. Homogeneous systems present processes where the catalyst and reactants are in a solution [34], while in heterogeneous processes, the catalyst is supported on a substrate [19].

GNPs can be used for selective oxidation (particularly for carbon monoxide [35]) or, in some cases, for reduction, for pollution control applications, automobile catalysts, and purification of hydrogen streams used for fuel cells. Using other noble metals, such as platinum, in combination with GNPs, may even increase their catalytic activity, promoting the use of GNPs in the Automotive industry [34] and for the removal of carbon monoxide, for example, in air purification [35].

3.2.2 Electronics

In Electronics, printable GNPs may be used for radio frequency identification (RFID) tags [36] and flexible electronics in flat panel displays, low cost sensors and other disposable electronic devices [37]. In electronic chips, GNPs are used for connections for resistors, conductors, and other elements through additive printing. Current electronics production uses a series of additive and subtractive manufacturing methods, which are complicated, time-consuming and environmentally unfriendly. 3D printing of electronics allows for a simplified and direct fabrication of electrical circuits onto different types of substrates [38] (Fig. 3.7).

The appeal of using GNPs in printing inks is in their good thermal and oxidation stability, with high electrical conductivity [39]. The disadvantage is their high cost, when compared to other metal nanoparticle inks, and the subsequent low-cost effectiveness when used in large scale production of cheaper printed electronics [38]. Other advantages are the tunable optical properties of GNPs, which are useful for specific applications and gives the ability to sinter the printed particles with intense pulsed light (IPL) or lasers [39]. The oxidation resistance of gold enables the printing of devices to be used in harsh conditions for metals, such as for electrochemical sensors and flexible or wearable devices in contact with salty mediums like human sweat.

Fig. 3.7 A photo and SEM image of inkjet-printed films, patterns and electrode arrays with a GNP ink, adapted from [36]

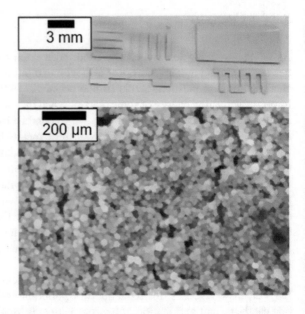

GNP printable inks are being used for conductive patterns, electrode and micro-electrode arrays, biosensing applications, flexible electrodes and wearable sensors [38, 40, 41].

3.2.3   Sensors

The intense colours and optical properties exhibited by GNPs makes them suitable for a number of various sensors. Gas sensors with GNPs, with selective detection of pollutant gases, which can detect CO and NOx, are used for monitoring air quality. Colour change sensors are used for monitoring components of body liquids [34]. Colorimetric sensors are used for identification of food contaminated with unwanted DNA, such as pork in a beef meatball [42]. These sensors can also be used for the detection of other proteins, contaminants and molecules. Examples include the detection of harmful anions in contaminated water [43], detection of biomolecular interactions [44], food safety [45], and a number of other chemical and biological sensing applications [46], such as sensors for detecting mercury [47].

The principle of sensor operation is a visible optical signal of detection in the event when a substance recognition or a binding event occurs. The visible signal is usually a colour change, as the plasmon resonance of GNPs changes when substances bind to their surfaces. The colour change is advantageous, as it allows for an easy interpretation of results.

GNPs are some of the most stable nanoparticles in sensors, with strong adsorption, which gives them the ability to be functionalised for selective testing. With their small sizes, the functionalisation is possible at nearly the molecular level, which gives them ultrasensitive detection capabilities [43].

3.2.4   Biomedical

A large number of studies and research activities focus on the use of GNPs for medical and biological applications. This field of research is populated by studies for photodynamic therapy, targeted delivery of therapeutic substances and various diagnostic methods. Photodynamic therapy uses GNPs that absorb near infrared light and produce heat in order to destroy tumour cells [48]. GNPs are suggested as effective drug delivery agents, able to release their payload at specific sites when subjected to an external or internal stimulus [49]. In diagnostics, GNPs are used to detect cancer and other diseases [50].

Lateral flow immunoassays (LFIA) are devices for diagnostics, which could also be described in the section of using GNPs in sensors, as they have the same principle of operation when the substance of interest is detected—a visible optical signal of detection. LFIA kits employ GNPs as the carrier for the analyte in rapid point-of-care tests, enabling specific and sensitive detection of pathogens causing human and veterinary diseases. The functionalisation capabilities of GNPs in LFIA tests make them useful for the detection of diverse viruses, bacteria, or other substances of interest. The home pregnancy test represents a common household example of LFIA tests. Rapid LFIA tests are especially useful for preventing or reducing the spread of global viruses, such as in the pandemic of the SARS-CoV-2 coronavirus outbreak [51] (Fig. 3.8). Refer to the introduction chapter for a more detailed description on the functioning of the LFIA test.

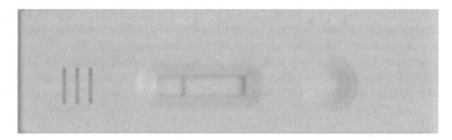

Fig. 3.8 Photo of a prototype LFIA rapid test developed for detecting infection with COVID-19 [51]. The GNPs contained in the test strip have coloured the control line red, no line is evident on the virus detection line

Fig. 3.9 24 K nano gold
anti-ageing skin care cream
from orogold cosmetics [55]

3.2.5   Cosmetics and Optics

GNPs are being used as ingredients in personal care items [52], such as skin care creams, sunscreens, makeup, soaps, lotions, shampoos, toothpastes, lubricants, as well as in decorative inks, glasses for optical limiting applications [53] and luxury products (Fig. 3.9). Toxicity and the effects of GNPs from personal care products on humans, as well as the environmental effects of GNPs, are the main points of investigation and concern in this field. As these products have a therapeutic effect on the skin, being used for treating conditions such as skin dryness, dark spots, uneven complexion, hyperpigmentation, hair damage and others, they are confined in an area between pharmaceutics and personal care products [54]. These effects are possible due to the GNPs interacting with the skin barrier, enhancing the delivery of active agents in the cream and improving skin permeability for active agents with a high molecular weight. GNPs are considered as promising candidates for skin immunisation and optimising transdermal delivery systems [16].

Regulations for the use of nanoparticles in cosmetics are not entirely explicit, but are progressing as more insight is collected on their use in these products, as they present one of the fastest growing segments of this industry [52, 54, 56].

In optics, the visible light absorption and scattering effects provided by SPR enables the use of GNPs in glasses as light filters, changing the refractive index and the absorption coefficient of the glass [53]. Application examples of GNPs on glass are spectrally selective coatings, such as elimination of blue light emission from display monitors, blocking UV light and for decorative applications (Fig. 3.10).

3.2.6   Microscopy

The optical properties of GNPs also make them useful in optical imaging, dark-field microscopy, surface-enhanced Raman spectroscopy (SERS), plasmon enhanced fluorescence (PEF) and two-photon or multi-photon imaging for biological imaging

Fig. 3.10 Representation of eyeglasses with a coating of GNPs absorbing the green and blue lights of the visible light spectrum, as well as UV light

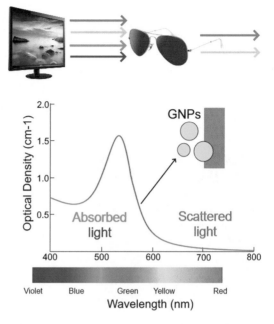

applications, in order to improve resolution and sensitivity [57]. The high density of GNPs makes them useful as dense markers in transmission electron microscopy (TEM), where they are used as probes to visualise and locate proteins by exploiting the binding of antibody-GNP conjugates [58]. The technique can be modified to provide identification of other molecules and biomolecules with the proper functionalisation of the used GNPs. GNP probes for TEM provide a straightforward signal that is easily noticeable in biological and inorganic samples.

3.3 Synthesis Techniques

The production or synthesis methods of GNPs are commonly divided into two main categories, as the top-down and bottom-up approaches. Both approaches include chemical, physical and biological synthesis methods, or a combination of these methods. The top-down approaches produce GNPs by subtracting material from bulk gold into small nanosized constituents with a strong attack force, such as ion irradiation in air or arc discharge in water [59]. Other top-down methods include mechanical milling, laser ablation, etching, sputtering and electro-explosion [24]. In bottom-up approaches the GNPs are formed from atomic or molecular components. With this approach GNPs are synthesised by chemical reduction of Au salts, electrochemical pathways and decomposition of organometallic compounds [59]. Bottom-up methods include chemical vapour deposition, solvothermal and hydrothermal

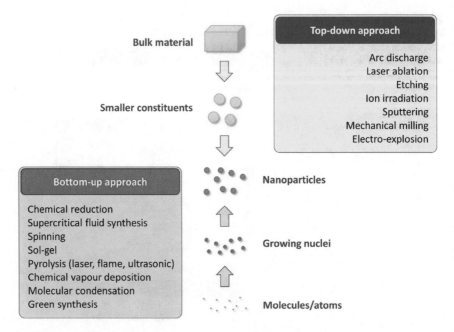

Fig. 3.11 Top-down and bottom-up approaches in GNP synthesis

methods, pyrolysis, sol–gel, templating and reverse micelle methods. The starting materials for producing GNPs in both approaches are from bulk material, gold seeds, gold salts and various biological extracts [17] (Fig. 3.11).

Although GNPs have a long history of research, the production of small and stabile nanoparticles with the desired size, shape and monodispersity remains a key challenge [14, 17]. In top-down approaches it is difficult to control the size and shapes of GNPs and to obtain a narrow size distribution, due to the limitations of these methods, which effects their physical and chemical properties greatly [17, 59]. In contrast, the bottom-up approach presents synthesis methods with more controllable production of various sizes and shapes of GNPs. These methods generally follow the same route for GNP formation in two steps, nucleation or formation of seeds, and growth of these nuclei into GNPs. Gold atoms, ions and clusters are formed by reduction, which then collide and bond into nuclei that grow into GNPs under the conditions of the synthesis. The growth conditions are the parameters of the given synthesis, such as the chemical and stabilising agents present, the pH level, temperature, and the physical or biological parameters and limitations (pressure, energy, time, etc.).

The formation of spherical GNPs requires an isotropic growth on the surface of the Au nuclei, while the controlled synthesis of different shapes is determined by constraining the growth into a single or several directions of interest with surfactants or other physical and biological restrictions for growth [17, 59] (Fig. 3.12). A greater number of bottom-up methods have been researched than top-down methods, as a

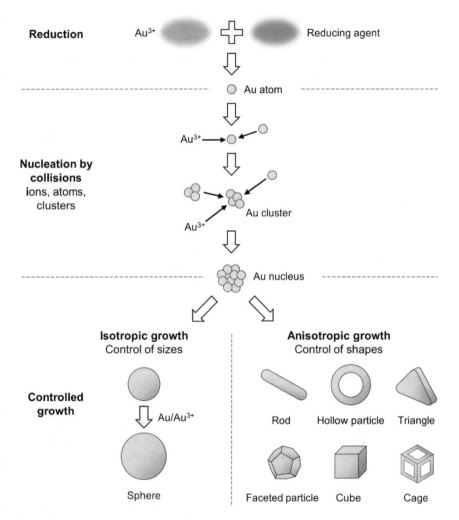

Fig. 3.12 Formation mechanism of GNPs with various particle sizes and shapes by the chemical reduction method synthesis, adapted from [59]

result of the more favourable conditions of controlling the nanoparticles' growth and producing the desired shapes and sizes of GNPs.

3.3.1 Top-Down Approaches

In the following section some top-down approaches will be described, for a general overview of the capabilities of these methods and the types of GNPs that are possible to be produced with this approach (Fig. 3.13).

Arc discharge

With arc discharge, two gold electrodes are submerged in water and subjected to an electric current. The temperature between the electrodes can reach thousands of degrees Celsius, vaporising the Au wires in the process. The Au vapours condense more easily in the medium than in air, producing GNPs. The method can produce spherical particles with sizes of about 15–30 nm [60].

Laser ablation

Laser ablation produces GNPs from bulk gold in a medium, with a powerful laser hitting the target material. Due to the high energy, the target material vaporises and results in the formation of GNPs in the medium. The particle sizes vary according to the wavelength of the laser, the exposure of the material to the pulsed laser and the surrounding medium with stabilisers [24]. Fairly small spherical GNPs can be produced with this method, with sizes ranging from 1 to 20 nm [24, 61]. Laser generated GNPs have a positive surface charge, which is an advantage compared to chemical routes. The charge causes an electrostatic stabilisation of the nanoparticles without the need for additional stabilising agents [62].

High-energy ball milling

With high-energy ball milling a powder mixture of the desired material is placed in the ball mill. The high energy impacts of the balls refine the powder into smaller constituents. Surface and interface contamination of the produced nanoparticles and size control is a major concern in this type of production method. Various sizes of a material may be produced in one batch, ranging from sub-micron to several tens of microns in diameter. With optimal milling, sizes of a few 10 nm may be obtained, limited by the grain size of the material milled [63].

Condensation with inert gas

In this process, bulk gold is vaporised in a vacuum chamber and then supercooled with inert gas. The gold vapours condense into GNPs, that are transported by the gas onto a substrate, or are studied in-situ. The method can produce very small particles, with sizes of 1 nm [64].

Lithography

In lithography, a focused beam of light, electrons or ions etches the target material, with or without a mask or template. in this method, arrays of nanostructures are produced simultaneously. In maskless lithography, nanopatterns are written by the beam for producing the desired nanostructures [24]. This method can produce GNPs

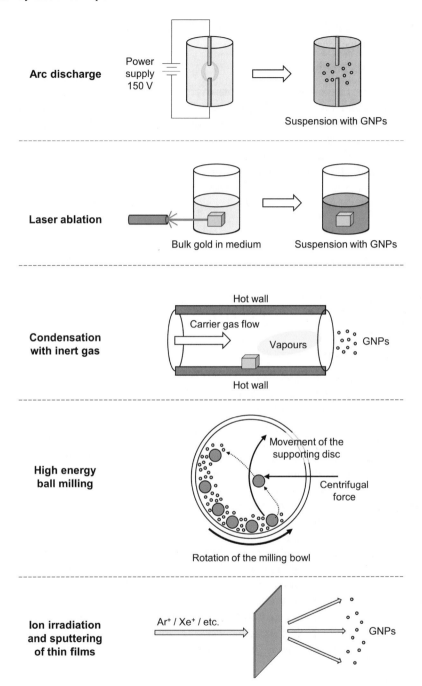

Fig. 3.13 Schematic representation of some top-down approach methods for synthesising GNPs

with varying sizes and shapes, depending on the lithography pattern. Sizes from 2 to 4.5 nm have also been reported [65].

Sputtering techniques

Gold thin films were irradiated with ions of Ne, Ar, Kr and Xe at varying dose rates and acceleration voltages in a Transmission Electron Microscope, sputtering GNPs from the back of the Au thin film with a thickness of 62 nm [66]. The resulting GNPs were small, up to 4 nm in size, and were collected on a carbon film inside the TEM vacuum with no additional stabilisation.

In general, sputtering techniques involve bombarding the target materials with high-energy particles, such as ions, plasma and gases. The bombardment ejects atom clusters from the target material and deposits them on a substrate [24].

3.3.2   Bottom-Up Approaches

Bottom-up approaches are more numerous than top-down approaches, with a wide range of possibilities for production of GNPs with different shapes and size ranges (Fig. 3.14).

Wet-chemical routes

Chemical reduction methods involve two steps for GNP formation, nucleation and growth. In situ syntheses include these mechanisms in the same process, while seed-growth syntheses include various steps to achieve GNP formation [67]. With in situ methods, reduction of gold containing precursor salts (most often $HAuCl_4$) is carried out by various reduction agents, from borohydrides (sodium borohydride), bromides (tetraoctylammonium bromide—TOAB, cetyltrimethylammonium bromide—CTAB), aminoboranes, hydrazine, formaldehyde, ascorbate, hydroxylamine, citric and oxalic acids such as sodium citrate, saturated and unsaturated alcohols, polyols, thiols, sugars, hydrogen peroxide, sulfites, carbon monoxide, hydrogen, acetylene and others [67–69]. Usually, a surfactant is also used in situ, for stabilisation and functionalisation of the formed GNPs.

One of the most common routes used for producing GNPs is the Turkevich method, a chemical route developed by Turkevich and coworkers in the 1950s [70]. The method uses small quantities of hot chloroauric acid ($HAuCl_4$) which reacts with small quantities of a sodium citrate solution. The sodium citrate acts as a reducing and stabilising agent. This method was most notably improved upon by Frens, by producing almost spherical particles over a tunable range of sizes [71]. Most of the modern chemical methods are modifications of this technique, with variations regarding the parameters and constituents used for reduction of gold containing salts into GNPs, suspended in a medium.

Another highly used synthesis technique, the Brust-Schiffrin method, uses a two phase reaction to produce small 1.5–5.2 nm GNPs in organic solvents, such as toluene [49]. In this method, tetraoctylammonium bromide (TOAB) acts as a phase transfer agent to transfer the gold salt from its aqueous solution to an organic solvent. Sodium borohydride, $NaBH_4$ then reduces the gold ions in the organic solution. The method and its variations are also able to produce highly stable thiol functionalised GNPs, in organic solutions [72].

Although GNP production with chemical routes has been being developed for decades, precision control of the particle size and size distribution is still an important challenge, due to batch-to-batch variation in a local temperature gradient, the efficiency of reagent mixing and the resulting local concentration gradient [71, 73]. These inconsistencies in production make the scale-up of GNP synthesis complicated, causing high precision GNPs to be available only in small quantities with high prices, even above $10,000 per gram [73].

An advantage of the chemical routes is presented by the highly tunable sizes and shapes that are possible to be produced with these methods, as they allow many available reactants and surfactants to be used for chemical reduction from the precursor, and particle growth along with functionalisation, for use in specific applications. This promotes a widespread adoption and development of these techniques, evidenced by the large number of publications in the literature involving chemical synthesis and use cases of GNPs produced by chemical methods.

Other types of chemical route methods are usually derived from these fundamental synthesis methods, also involving other physical or biological methods for promoting or constraining particle growth, or otherwise controlling the particle formation parameters, morphology and structure.

Seed-Mediated Growth

This method is preferred for the production of GNPs in shapes other than spherical, such as rods, cubes, tubes, stars, flowers, dendritic or polyhedral nanostructures. Arranging the surface atoms and shapes of the GNPs allows for tuning of the SPR band and optical properties, increasing the possibilities of use for these nanoparticles. The seed particles are produced with reduction of gold salts, and then added to a gold salt solution in the presence of a weak reducing agent such as ascorbic acid, along with a surfactant which accelerates the anisotropic growth of the GNPs in a specific direction. A widely used surfactant for this purpose is cetyltrimethylammonium bromide (CTAB) [14, 49, 74, 75].

Sonolysis

In sonolysis, the reduction of a gold containing precursor ($HAuCl_4$) in a solution with a stabiliser, such as sodium dodecyl sulfate or glucose, is enabled by subjecting this solution to an ultrasound [68]. The high frequency ultrasonic waves irradiate the solution, and high pressure collisions occur between the solution monomers for nanoparticle formation [75]. Changing the synthesis ultrasonic power and stabiliser gold salt concentration produces GNPs with a wide size range, up to a few 10 nm in spherical and irregular shapes [76], or gold nanoribbons with a width from 30 to 50 nm [77].

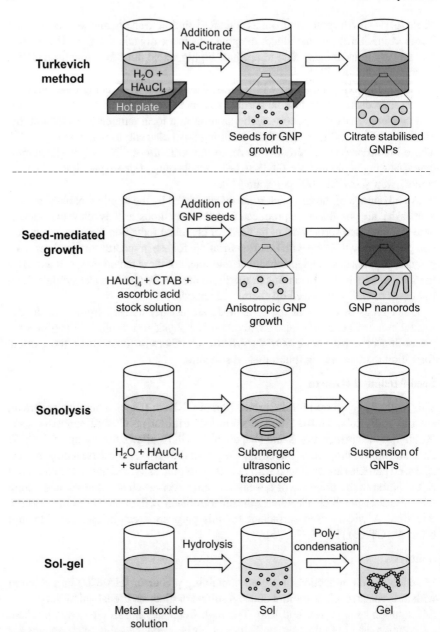

Fig. 3.14 Schematic representation of some wet-chemical bottom-up approach methods for synthesising GNPs

Sol–gel

Sol–gel forms a colloidal solution (sol) from precursor monomers in solution, which form gradually into a three-dimensional solid network structure (gel). The precursor is a metal alkoxide or chloride. In the process hydrolysis of the metal oxide is performed in water, or with the assistance of alcohol to form a sol. Afterwards, condensation takes place, resulting in an increase in the solvent viscosity, to form porous structures that are left to age. As polycondensation continues, the nanoparticle structure and its properties change, while the porosity increases, increasing the distance between the colloid particles in the solution. Finally, drying may remove the organic solvents and water from the gel [24]. Sol–gel is a cheap, low-temperature method, used for processing and production of ceramics and thin films with GNPs with sizes below 20 nm [78].

3.3.3 Aerosol-Based Synthesis Methods

In comparison with wet chemical processes, aerosol-based synthesis methods produce GNPs from aerosols from a gas or liquid phase. In liquid phase aerosol methods, the solution, with a specific composition, is atomised with electrospray, ultrasound, hydraulic or pneumatic atomisers, to form droplets of various sizes, about 1–10 microns in diameter. The fine droplets then evaporate and crystallise to solid nanoparticles. In gas phase aerosol methods, the particles are generated by nucleation of the gas phase and growth of the nuclei by condensation and coagulation. Liquid atomisation is achieved with nebulisation or electrohydrodynamic atomisation, while gas atomisation is achieved with flame, furnace, plasma, or spark discharge techniques [79].

Aerosol-based synthesis methods include flame, ultrasonic and laser spray pyrolysis, plasma synthesis, spray drying and spark discharge (Fig. 3.15). They are low cost and versatile methods with good control over the particle sizes and shapes, making them a popular choice for synthesis in recent years. These methods are being adopted on a commercial level, as they allow the collection of GNPs in a one-step process, with low waste and a minimal environmental impact [79, 80].

As the principles of GNP formation with these methods are very similar, only a few of these methods are described in the following section.

Flame spray pyrolysis

Flame spray pyrolysis is a synthesis method that forms GNPs from an aerosol in the gaseous phase with high temperatures in the flame. The technique produces nanostructures from a sprayed liquid precursor, which evaporates or combusts, decomposes or oxidates, and forms nuclei in the flame, which grow into solid nanoparticles by coagulation, coalescence, or sintering and agglomeration [79]. The sizes, shapes and rate of agglomeration (hard and then soft) depend on the liquid precursor and oxidation agent delivery rate, as well as the various nozzle parameters, residence time

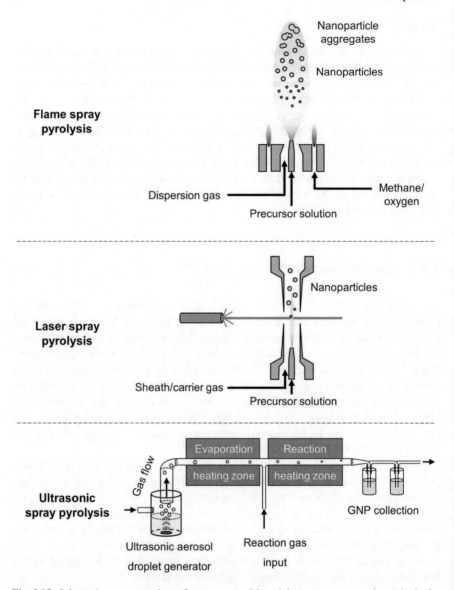

Fig. 3.15 Schematic representation of some aerosol-based bottom-up approach methods for synthesising GNPs

of the particles and temperature of the flame. The particles may be collected on a substrate for thin film deposition. GNPs can be obtained readily in oxygen rich flames with particle sizes below 50 nm, depending on the parameter conditions. For narrow particle size distributions, steep homogeneous cooling gradients are necessary, due to the inherent particle growth and agglomeration. For non-noble metals reducing conditions are required [80].

Laser spray pyrolysis

With laser pyrolysis, a high-powered laser is used to vaporise the dispersed precursor, which is followed by nucleation and condensation for particle growth, or a chemical reaction or decomposition to form the nanoparticles. As the precursor vapours cool down in the carrier gas stream, further condensation may produce clusters and nanoparticle aggregates [81].

This technique allows for the use of liquid, gaseous and aerosol precursors for the production of nanoparticles. By varying the pulse energy and frequency of the laser, it is possible to produce various metal compositions [79]. GNPs with primary sizes of about 10 nm and aggregates of 45 nm were prepared with the laser technique under a continuous flow of nitrogen gas [81].

The nanoparticle production yield with laser pyrolysis is often low, which, in addition to expensive laser investments and the high costs of operation, makes this method less economically convenient in other extents than on an industrial scale [79, 80].

Ultrasonic spray pyrolysis

With ultrasonic spray pyrolysis, the precursor solution is dispersed into aerosol droplets with ultrasound, and transported into a reaction furnace with an inert gas. The droplets then evaporate, and the dry particles decompose at high temperatures. A reaction gas may be included for formation of nanoparticles from the dry particles by reduction. For GNP production, using an aqueous solution with gold chloride or acetate, with nitrogen as the carrier gas and hydrogen as the reaction gas, is favourable for nanoparticle formation. The nanoparticles may be collected in a solution with suitable stabilisers.

Depending on the precursor volatility and process parameters, the GNP formation can be from the gaseous or liquid phase in the reaction furnace. Changing the process parameters of gold concentration in the precursor solution, precursor type, gas flows and reaction temperatures, yields GNPs in sizes from 10 nm up to submicron level. The size distribution may be broad, from 10 up to 300 nm, or narrow, from about 20 to 50 nm, depending on the parameter optimisation. Research is focused mostly on the production of spherical shapes with this method. Similar to other aerosol-based synthesis methods, the GNPs should be cooled rapidly after formation, in order to avoid further agglomeration or coalescence and particle growth.

The method offers a high flexibility in precursor composition and parameter selection. As such, several different types of nanoparticles have been produced by this method, solid or hollow nanoparticles, core–shell, ball-in-ball structures, yolk–shell-structures and multicomponent composite particles, as well as for thin film deposition. It is possible to produce mesoporous particles of various materials and chemical compounds by confining the sol–gel chemistry to single aerosol droplets with micron sizes [82]. This method is also considered to be relatively cost effective and up scaled easily to an industrial level.

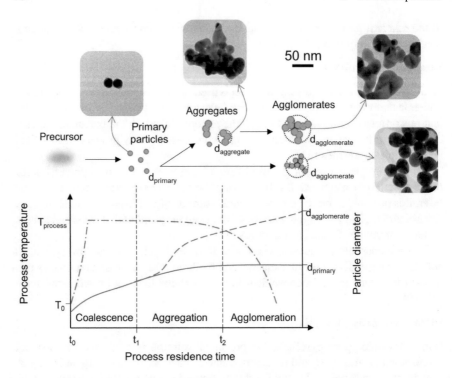

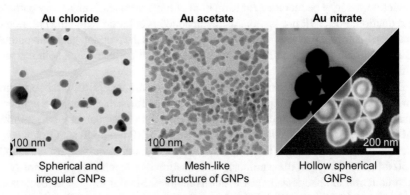

Fig. 3.16 GNP growth by the formation of primary particles, aggregation and agglomeration, depending on the chemical and physical process conditions and residence time in each formation stage. With optimal conditions, uniform shapes and sizes are possible to be produced with ultrasonic spray pyrolysis. Examples of the GNP types produced by different starting precursors are shown

In our research of synthesising nanoparticles with ultrasonic spray pyrolysis, we have produced GNPs from gold chloride [83], gold acetate [84], gold nitrate [26] and dissolved gold scrap [85, 86], along with synthesis of core–shell particles with a non-continuous shell layer from TiO_2/Au [87, 88] and Fe/Au [89, 90]. In producing single component GNPs, the formed nanoparticles show that the gaseous and liquid phases can also be present in the furnace simultaneously. The precursors of gold chloride, acetate and nitrate are acidic and volatile, requiring great control over the synthesis process in order to produce AuNPs of the desired shapes and sizes, due to the mixture of gas and solid phases present in the reaction tube, as well as the inherent physical and chemical phenomena that are being performed inside the tube furnace (collisions, coalescence, aggregation, agglomeration, etc.). Particle formations in these conditions are determined by the droplet-to-particle and gas-to-particle mechanisms, with finely determined synthesis parameters of precursor concentration, gas flow and reaction temperatures, in order to exploit the different formation mechanisms from the two phases and to produce GNPs with controlled sizes and shapes [91].

One of the factors for GNP growth is the residence time of the aerosol particles in the reaction furnace, which determines the length of time when the particles are in each formation stage. The final morphology is highly dependent on these processes, as seen in Fig. 3.16. The residence time can be divided into three stages: Coalescence (formation of primary particles), aggregation (formation of hard agglomerates), or agglomeration (formation of soft agglomerates). The length of the aggregation stage is the main condition for producing uniform, unaggregated particles. The process chemical and physical conditions are important in this stage, such as the volatility of the precursor, which determines the ratio of the growth by condensation and growth by particle coagulation. Two growth routes are shown in Fig. 3.16, depending on the length of the aggregation stage. In our experiments, the GNPs were more uniform and spherical when the residence time was short, showing agreement with the formation model. An important factor for producing such GNPs is also their rapid cooling after formation. In ultrasonic spray pyrolysis, the GNPs are collected in a liquid medium, which should be cooled to control particle agglomeration after collection.

Further research of the GNPs produced by ultrasonic spray pyrolysis was conducted by using them in stability and nanoparticle surface investigations [92], in cytotoxicity studies for biomedical applications [85, 86, 93–95], as biomarkers in rapid LFIA tests for diagnosing Coronavirus COVID-19 [51], as a filler component in dental resins [96], as printing inks for electronic devices and sensor arrays [97, 98], in skin care cosmetic products and for coatings on eyewear products.

3.3.4 Biological-Based Synthesis Methods/Green Methods

The chemical and physical methods for GNP production have several advantages, such as a high yield, low costs and reproducible results in sizes and shapes. Some disadvantages, such as high energy inputs, precursor contamination, waste, byproducts, solvent toxicity and the use of harsh chemicals, may be overcome with the use of biological based or green types of synthesis methods. These methods offer economical, energy efficient, clean and environmentally harmless techniques for GNP production with the use of microorganisms. These vary from simple bacterial cells to complex prokaryotes and eukaryotes that produce nanomaterials in vivo. The wide range of green techniques for GNP production include plant-based compounds and derivatives, bacteria, plants, algae, fungi, yeast and viruses [17, 49, 68, 72, 99] (Fig. 3.17).

The reduction of metal ions in biological agents is carried out under mild temperature and pressure conditions (room temperature, atmospheric pressure) and physiological pH, with small quantities of organic solvents. The metabolites which carry out the reduction (proteins, fatty acids, sugars, enzymes and phenolic compounds) are also important for the stability of the produced nanomaterials [72]. GNPs produced with biological synthesis methods are reportedly more stable than similar nanoparticles produced with other methods. The disadvantages of these methods are low yields and difficulties in handling for some cases, such as with bacteria [17].

Bacteria can carry out the synthesis of GNPs with enzymes and biomolecules intercellularly or extracellularly [49]. When produced inside the cell, additional steps are needed for purification of the GNPs. Fungi secrete several metabolites and extracellular enzymes used for reduction into GNPs. Plant extracts are used commonly as well, which hold the bio-components of the plant, such as flavonoids, phytosterols, quinones and others. These are important for GNP synthesis due to their functional groups, which promote the reduction and stabilisation of GNPs. Biomass from marine and freshwater algae contain hydroxyl and carbonyl groups, which carry out the reduction and stabilisation of the GNPs. Biomolecules, or molecules produced by living organisms, such as amino acids, nucleic acids, carbohydrates and lipids, are also used for reduction of Au ions into GNPs [49, 72]. Even whole plants, which hyperaccumulate metals, may be used as biological factories for GNP synthesis during in vivo phytosynthesis [100].

In general, the sizes of GNPs produced with biomaterials range from a few nm to several tens of nm, usually in the range of 5–50 nm. The shapes are primarily spherical. Other shapes of GNPs, such as triangular, rods, pentagonal, hexagonal, are also possible to be produced, with sizes up to about 100 nm or more [68, 99].

Extracellular synthesis provides an advantage to intracellular synthesis, as additional steps may be avoided for the procurement of pure GNPs. Some of these methods take quite a long time for GNP production, as raising the necessary agents to perform the reactions, such as bacteria or algae, may take hours or several days. In some methods, the extraction of the necessary biomolecules presents a highly precise and difficult step in the synthesis. In some cases, it is difficult to identify the reducing agent responsible for GNP formation, as there are many organic components present in the synthesis procedure. However, these may also aid in the reactions and stabilisation or functionalisation of the produced GNPs [49, 99].

Biological based synthesis methods have numerous options for tuning the GNP properties, and are considered to be ecologically friendly and reproducible. The high

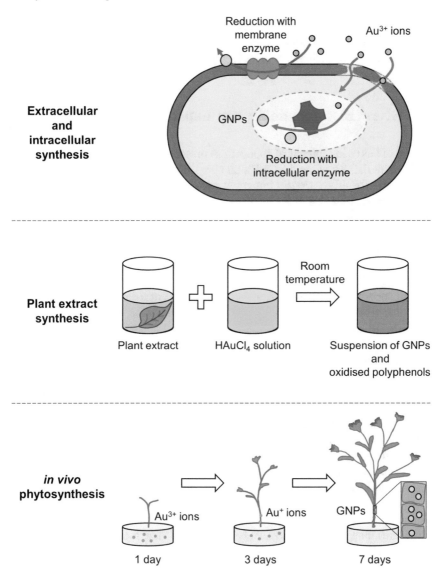

Fig. 3.17 Schematic representation of some biological based synthesis methods for producing GNPs

stability and the biological and immunological behaviour of the GNPs produced in biological environments makes these methods attractive for research in biomedical applications, for targeted drug delivery, biodetection, medical imaging, for photothermal, hyperthermal and gene therapy. As such, research continues for

explaining the different synthesis mechanisms, and determining the reduction capabilities of various biomaterials, before using them in scaled up productions for GNPs
to be used in these applications [49, 68, 99].

3.4   GNPs' Biocompatibility and Impact on Human Health

The use of GNPs in biomedical applications presents one of the main research areas
for this material. Knowing the GNP's biocompatibility and behaviour in tissues and
organisms is, therefore, a highly important area of study. As this topic presents a broad
field of investigation due to the many uses of GNPs in biomedicine and other areas,
only a brief overview of the currently known observations of GNPs' functioning and
their effects on health is given here.

Nanomaterials are rapidly emerging tools for applications in different fields.
This raises concerns on the safety, toxicity and biocompatibility while using them,
disposing of them and during their production. The biocompatibility and toxicity of
these materials is known to be dependent on the type of material, size, surface area,
functional groups, concentration, charge and dosage [101].

Due to the novelty of products engineered with nanomaterials, there are currently
no harmonised Standards for evaluating their harmful effects towards the user. Regu-
latory organisations are, thus, examining the applicability of established testing
frameworks, as well as evaluating and compiling new legislation for the production, use and disposal of nanomaterials. In order to understand the toxicity effects of
nanomaterials, they are being tested in in vitro, in vivo and in silico assays.

Nanotoxicology is a moderately new branch of toxicology that addresses the gap
in the knowledge of toxicity induced by nanomaterials. The interactions of nanomaterials with living tissues and their behaviour are highly susceptible to their physicochemical state. A single variation in the nanomaterial property can lead to a complete
change in its behaviour and its effect on living cells. Nanomaterials have been known
to induce asthma, dermatitis, rhinitis, lung diseases and contaminations, tuberculosis,
respiratory embolism, immune system illnesses and others. The results of toxicology
assays are, thus, critical for researchers, to direct their focus of development on
nanomaterials that are relatively harmless to the population and the environment
[102].

Bulk gold is proven to be biologically benign, stable and inert. These properties
also apply to micron sized gold particles, while the toxicity effects of the nanoform
of this material are still under evaluation. Many studies consider GNPs to be biocompatible with little cytotoxic effects (toxicity towards cells), while others contradict
this statement [103–105]. As their physicochemical properties affect their behaviour
greatly, GNPs of different synthesis methods, shapes, sizes and surface chemistries
are being studied for their inherent beneficial or detrimental characteristics. Additionally, the GNP's effects are also highly specific to different tissues, organs and the
methodology of administering them (inhalation, absorption through skin, injection,

etc.). Due to these factors, the cytotoxicity and biocompatibility of GNPs must be evaluated on an individual and specific application basis [105].

The sizes of GNPs are studied extensively, as this parameter determines the cytotoxicity and cell uptake of GNPs greatly. GNPs enter cells mainly by clathrin-mediated endocytosis, and exit cells by exocytosis [105]. Smaller GNPs are considered to internalise in cells more effectively as compared to larger particles, which need to interact with more receptors on the cells' surfaces. GNP cell uptake is considered as an energy dependent process [20]. A greater number of studies of GNPs consider smaller sizes up to about 20 nm to be more cytotoxic [102], while some others conflict with this conclusion, stating larger sizes are more toxic, or that size has no effect on cytotoxicity [101, 105]. This is possibly a result of the different synthesis methods, surface chemistries and assay conditions employed for the various investigations.

GNPs with sizes of 10 and 50 nm were used in studying their effects on human dendritic cells in vitro. It was shown that the 10 nm sized GNPs interfered with the maturation and function of these immune cells, causing their inability to trigger an antitumour response, and could be interpreted as adverse in tumour therapy. After cell internalisation of the GNPs, 50 nm sized particles appeared more clustered, while the 10 nm particles were more dispersed, and were internalised in greater numbers. The smaller GNPs had a weak pro-apoptotic effect, causing some cell death, while the 50 nm GNPs had no cytotoxic effect. The size dependent GNP effects could be associated with different internalisation mechanisms, accumulation levels and distribution within the cells [20] (Fig. 3.18).

Investigations on the cytotoxicity of differently shaped GNPs show that spherical shapes are the least detrimental, while rod like, stars and elongated shapes seem to be the most toxic [101, 102, 105]. A higher number of atoms at angles and edges, characteristic of these shapes, may cause stronger interactions with biomolecules and toxicity [105]. Spherical shapes are also internalised more easily than other shapes [101, 102].

The cytotoxicity of GNPs can be controlled by surface charge and surfactants, which also prevent agglomeration in physiological and biological mediums, alter cellular uptake, and can be tuned to specific interactions between GNPs and cell surfaces and their responses [103, 105]. GNP surfactants are also used for drug delivery and imaging. Peptides, antibodies, PEG, citric acid, CTAB, nucleic acids and others have been used for GNPs with specific biomedical properties [105]. CTAB was shown to be highly toxic to human dermal fibroblasts in comparison to others [103]. Non-stabilised, sodium citrate, PEG and PVP stabilised GNPs with average sizes of 20 nm, produced by ultrasonic spray pyrolysis, were tested for cytotoxicity on mouse L929 fibroblast cells, B16F10 melanoma cells and human peripheral blood mononuclear cells (PBMNCs) [93]. The non-stabilised, sodium citrate and PEG stabilised GNPs were non-citotoxic to L929, B16F10 and PBMNC, while the PVP stabilised GNPs reduced the relative metabolic activity of L929 and B16F10 cells significantly in the chosen concentrations up to $100 \, \mu g/mL$. While PVP is considered biocompatible and used extensively in pharmaceutics, it can disrupt cell membranes and cause cell necrosis. However, it did not induce cytotoxicity in human PBMNC cells [93] (Fig. 3.19).

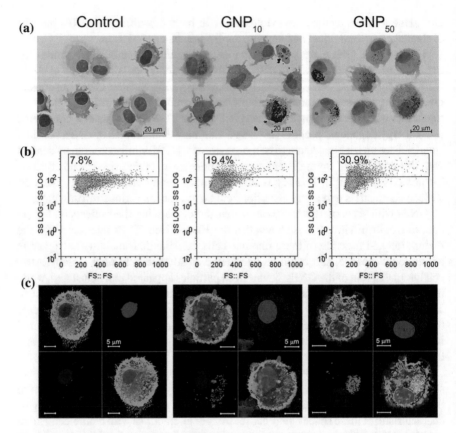

Fig. 3.18 Internalisation of 10 and 50 nm GNPs by human dendritic cells. Analysis **a** after staining the cells with May-Grunwald Giemsa (MGG), **b** by flow cytometry, **c** by confocal microscopy; the red dots represent GNPs, adapted from [20]

As a result of the numerous preparation methods and states of GNPs for biological assays and the diverse assay conditions, conflicting conclusions are available regarding GNP cytotoxicity and biocompatibility. The majority of studies claim that GNPs are biocompatible or non-cytotoxic, typically depending on their aforementioned features. There is also a lack of data correlation between in vitro and in vivo studies. More studies are needed to understand the possible adverse effects of GNPs and their fate inside the human body. The combined outcomes and conclusions from these studies form a broad overview in toxicity databases for GNPs, which are imperative for use of these materials on the larger scale, in biomedicine, products and industrial applications. A comprehensive awareness of GNPs' impact and safety on humans and the environment and predicting the effects of newly engineered products with this material, is, thus, not yet known completely. For this reason, the gaps are being filled with individual studies on a case-by-case basis, with variations on

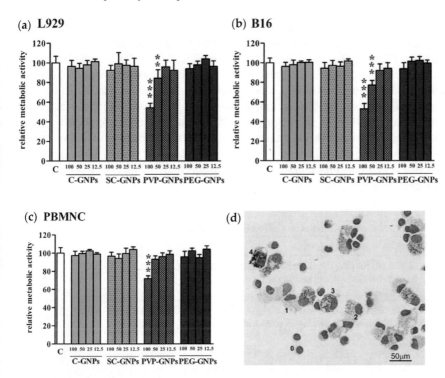

Fig. 3.19 The effect of bare, sodium citrate (SC), PVP and PEG stabilised GNPs with different concentrations up to 100 μg/mL on the metabolic activity of cells for 24 h in vitro. **a** Mouse L929 fibroblast cells, **b** B16F10 melanoma cells, **c** peripheral blood mononuclear cells (PBMNCs), **d** an example of the internalisation of GNPs in PBMNCs, adapted from [93]

GNP physicochemical properties, concentrations and doses, etc., which influence the interactions with biological media, cells, tissues and organs.

Several nanodrugs and nanopharmaceuticals have been approved in recent decades, while a smaller number of clinical trials for tumour treatments are currently underway [105]. As new discoveries are being made, the Regulations and knowledge bases will also be updated accordingly, with safety precautions and measures for minimising the health and environment concerns in the different uses of GNPs.

3.5 The Current State and Development of Gold Nanoparticles for Biomedical Applications

Using GNPs for biomedical purposes presents the greatest research interest for this material. They can assist in drug targeting, controlling drug release, increasing drug penetration ability [106], pathogen detection, in photodynamic and photothermal therapy. Cancer treatment is one of the most researched areas of GNP use [107]. They are also used for eliminating pathogenic microorganisms such as bacteria, fungi or viruses, for detection of Severe Acute Respiratory Syndrome (SARS), Ebola and the hepatitis C virus. They have an anticoagulant effect in blood plasma and can assist in controlling thrombotic disorders, while their inflammatory and antioxidant properties enable them to be used in treating neurodegenerative, chronic brain diseases associated with tauopathy, neuroinflammation and oxidative stress in the cortex and hippocampus [108]. As the list expands with new discoveries for GNPs, and as their safety, physicochemical properties and metabolic pathways are being studied as discussed in the previous section, a general agreement in medical research and practice is that Nanomedicine is likely to revolutionise the diagnosis and treatment of human diseases [106] (Table 3.2).

Table 3.2 Some clinical studies on GNPs from ClinicalTrials.gov [109]

GNP composition and application	Study status
Phase I trial of studying the side effects and best dose of TNF-bound colloidal gold in treating patients with advanced solid tumours	Completed 2006–2009
GNPs with silica-iron oxide shells for high-energy plasmonic photothermic burning of plaque in atherosclerosis treatment	Completed 2007–2016
AuroLase therapy: PEG-coated silica-gold nanoshells for near infrared light facilitated thermal ablation of head and neck cancer, lung tumours and prostate cancer	Some studies completed 2009–2023
Array of chemical nanosensors based on organically functionalised GNPs and carbon nanotubes for diagnosis of gastric lesions from exhaled breath and saliva	Completed 2011–2020
Evaluation of the safety, tolerability and pharmacokinetics of a gold nanocrystal suspension drug in healthy volunteers	Completed 2015–2016
Phase 1 study of GNPs attached with a peptide fragment related to insulin for treating Type 1 diabetes	Unknown 2016–2020
Nucleic acids on GNPs for drug delivery in glioblastoma multiforme or gliosarcoma treatment, early phase	Completed 2017–2020
GNPs conjugated to CD24 as specific biomarkers for early detection and treatment of salivary gland tumours	Completed 2018–2021
Mixture of GNPs and Ag nanoparticles suspended in 70% isopropyl alcohol for surface pretreatment of class II caries cavities before applying a tooth resin composite restoration [110]	Completed 2019–2021
T-cell priming peptides on GNPs in a vaccine against Dengue (naNO-DENGUE)	Estimated 2021–2022
T-cell priming peptides on GNPs in a vaccine against Coronavirus COVID-19 (naNO-COVID)	Estimated 2021–2022

3.5.1 Development of GNPs for Dental Applications

GNPs also have a role in the development of compelling applications in Dentistry, as described in the Introduction section. The antibacterial effect of GNPs and Ag nanoparticles was used against organisms like Streptococcus mutans (S. mutans), which cause tooth damage, breakdown of the enamel and caries [110]. For dental implants, GNPs can be used as osteogenic agents for promoting bone regeneration by accelerating cellular activity and growth, a feature also used in stem cell technology. Human bone marrow-derived mesenchymal stem cells (hMSCs) and human dental pulp stem cells (hDPSCs) were treated with GNPs with different functional groups, which were found to promote cell proliferation over osteogenic differentiation in hMSCs, and to improve the osteogenic ability of hDPSCs significantly. This type of tissue engineering may also be used for treating periodontal disease, where dental plaque, with its presence of microorganisms in a biofilm, causes loss of supporting tissue of the periodontium and the supporting alveolar bone, resulting in teeth mobility. GNPs show promising results in proliferation of stem cells for the reconstruction of tissues and bone, and as antibacterial agents for the eradication of harmful biofilms on teeth in periodontology [111].

The mechanical properties of dental resins may be improved by GNPs, without causing toxicity to the surrounding tissues and cells. Their use in dental adhesives inhibits Matrix metalloproteases (MMP) and enhances the flexural and tensile strength of the adhesive. In the adhesives, the GNPs act as obstacles for crack growth, improving flexural strength [111]. The addition of GNPs in polymethyl methacrylate (PMMA) and other polyacrylate composites is found to increase the flexural strength and thermal conductivity to almost double the value of pure PMMA [112].

Using ultrasonic spray pyrolysis, GNPs were prepared from gold (III) acetate (AuAc) and gold (III) chloride (AuCl), to be mixed with acrylic acid and acrylamide composites for dental applications. The AuCl prepared GNPs were spherical, with average diameters of 57.2 and 69.4 nm, while the AuAc prepared GNPs were ellipsoidal, with average diameters of 84.2 and 134.3 nm. SEM analysis showed a uniform distribution of GNPs in the polymer matrix. The densities of the GNP prepared composites were up to 40% higher as compared to pure composites. Compressive tests showed that the GNP composites had lower compressive strength, while their toughness increased. The results showed that GNPs do not improve the compressive strength of dental composites, however, they increase their toughness significantly [96] (Fig. 3.20).

3.6 Hybrid Gold Nanoparticles with Particles of Other Materials

The beneficial properties of GNPs can be enhanced further by combining the GNPs with other materials in various nanostructures. Multicomponent composite GNPs

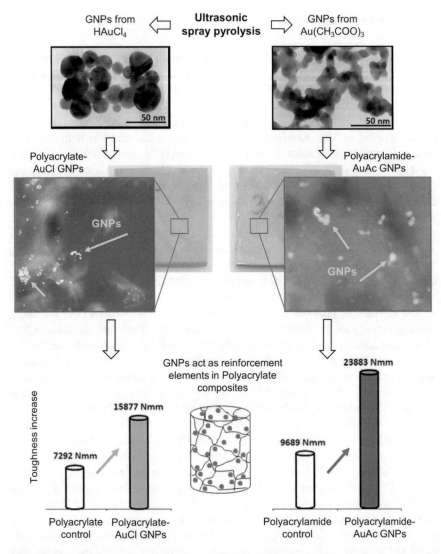

Fig. 3.20 Dental polyacrylate composites with GNPs, produced with ultrasonic spray pyrolysis. The GNPs do not improve the compressive strength of polyacrylate composites, but increase their toughness significantly [96]

with advantageous SPR functionalities may, thus, also be more catalytically active, magnetic, with enhanced thermal and chemical stability, even during repeated use or enhanced biocompatibility [13, 24, 89, 91]. The bimetallic or multicomponent GNPs consist of two or more materials, and are available as core–shell, ball-in-ball structures, yolk–shell structures, mesoporous, bimetallic mixture, alloys and multi-component composites (Fig. 3.21). A number of materials have been used for the

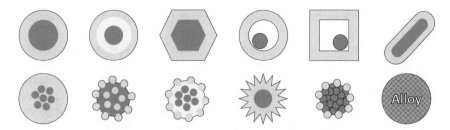

Fig. 3.21 Various GNP core–shell and hybrid structures with other materials

synthesis of hybrid GNPs, such as silica, graphene oxide, Ag, Cu, Pd, Pt, Zn, Fe, ZrO_2, CdS, Eu, Ti, Ni, Ru, Rh, and others [13, 24, 67, 91].

Core–shell structures are the most widely researched type of composite nanoparticles. They are composed of an inner material, termed the core, and a surface shell around the core. The typical shells on GNPs are silica, organic polymers, oxides and metals, as discussed in the stabilisation section. The properties obtained by adding a shell to the core material depend on the ratio of both component materials, as well as the shape, size and composition. Noble metal shells can also protect easily oxidisable or unstable cores for applications where these cores are needed, such as, for example, easily oxidisable magnetic nanoparticles [24].

In this section, hybrid GNPs with other materials are examined in more detail, along with the different processes used for synthesising these materials and their beneficial properties for use.

3.6.1 Core–Shell Bimetallic GNPs

The first GNP core–shell bimetallic particles appeared in the 1970s as Au core and Pd shell (Au@Pd) and Pd core and Au shell (Pd@Au). The Au@Pd core–shell nanoparticles were shown to be more catalytically active for hydrogenation reactions as compared to only Pd nanoparticles, due to their synergistic electronic effects [67]. Core–shell morphologies generally provide better catalytic activity, due to the synergistic effect of the metallic core–shell components [24]. Magnetic nanoparticles of Fe_3O_4 with an Au shell are interesting for biological and medical uses, as they provide a magnetic property, along with enhanced SPR, biocompatibility and possibilities for functionalisation, which are promising for diagnostic and therapeutic applications such as Magnetic Resonance Imaging (MRI) and hyperthermia—cancer therapy via heat treatment of cancer cells [67, 91]. Magnetic particles can be dispersed properly and delivered to their site of activation with an external magnetic field. Several magnetic particle cores are available, such as Fe oxides, NiO, Ni, Co and Mn_3O_4 [24]. Uncoated, these particles are not usually stable in biological or physiological environments. Obtaining an Au shell on these particles is, thus, an important course of action for achieving the desired combined properties of these particles for biomedical use.

Core–shell GNPs are usually synthesised by the reduction of several precursors in separate steps. The ratio of precursors in the solution determines the shell thickness, while the reduction potential is the determining factor for the core and shell formation. For example, the Au^{3+} ion precursor is easier to reduce than various transition metal cations, such as Cu^{2+} or Pd^{2+}. When a reducing agent is added to the precursor solution, such as $NaBH_4$, a core of Au particles is formed, followed by a shell of the lower reduction potential metals [67]. The thickness of the transition metal shell can be controlled with the initial precursor concentration of these metals [24]. In order to form a gold shell on other metal cores, the cores are usually synthesised first, and added to solution in which a gold shell can be formed either by growth on the cores, which act as seeds, or by replacing a core surfactant.

3.6.2   Iron Oxide-GNPs

The magnetic properties of iron oxide particles, beneficial for biomedical applications, may potentially be reduced by using a diamagnetic shell, such as Au or Ag in the core–shell structure. Despite this, the benefits of a gold shell (biocompatibility, functionalisation) for the desired use cases, have made this the material of choice for a lot of research in synthesising magnetic core–shell particles. The Au shell is usually not thick enough to suppress the magnetic properties of the core, such as coercivity or blocking temperature, while shielding the core from oxidation [91, 113].

Generally, the synthesis of the iron oxide core (Fe_2O_3, Fe_3O_4) is performed with wet-chemistry methods by the reduction of iron salts, such as ferric or ferrous chloride, in an aqueous solution with proper stabilisation. The formation of iron oxide cores is followed by reducing the $HAuCl_4$ in a solution already containing the synthesised iron oxide cores [114, 115]. Other synthesis methods are available, such as by exchanging the iron oxide surfactants with a gold layer [113, 116], or by using an intermediate layer such as silica [117]. The surfactant may also constrict the growth of the Au layer in specific facets, promoting anisotropic growth and the formation of structures other than spherical shells [118].

Single step synthesis approaches are also possible to produce iron oxide-Au core–shell nanoparticles. An attempt was made to produce Fe@Au core shell particles with ultrasonic spray pyrolysis, with the use of various combinations of iron containing salts and gold containing salts in the precursor solution [89]. The precursor solution was atomised into droplets with ultrasound and transported by gas into a tube furnace, where the reactions for particle formation were carried out. The resulting particles depend greatly on the type, concentration and ratio of the used precursor salts. Mainly, iron oxide cores covered with several small satellite GNPs were produced, not forming a whole Au shell [90]. Such a satellite-type layer provides a high gold specific area, with the possibility of new functionalisation options not available in GNPs, or as seeds for growing a complete Au layer and synthesising Fe@Au core–shell structures [118] (Fig. 3.22).

Combining the optical properties, biocompatibility, functionalisation and stability of GNPs with the magnetic properties of iron oxide particles, presents great potential for use in the biomedical area for magnetic, optical imaging, hyperthermia and photothermia therapy [118]. They have the potential for use for drug delivery, diagnosis and treatment of cancer, HIV and other diseases. The various synthesis methods

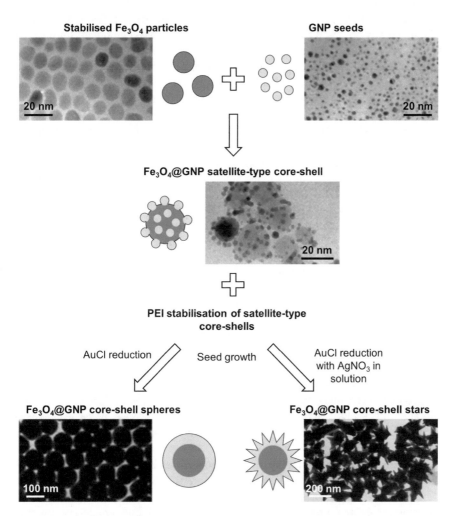

Fig. 3.22 Example of synthesising different Fe@Au core–shell particles. Satellite-type core-shells can be produced with amine stabilised Fe_3O_4 particles [119] or PEI stabilised Fe_3O_4 particles [120] with GNP seeds. After additional PEI stabilisation, the Fe@Au satellite-type core-shells can be processed further into core–shell spheres, or stars with additional GNP seed growth, adapted from [119, 119]

and combinations provide control over the core–shell particles' morphology and sizes with interesting properties. All of the methods present different advantages, disadvantages and challenges for core–shell particle production, which is still in its early developmental stage. The toxicity of these particles and their accumulation in the various tissues of the body still remain poorly understood, which is an important aspect for future research and development in this area [117].

With ultrasonic spray pyrolysis, research was conducted on the synthesis of $TiO_2@Au$ and $Fe_3O_4@Au$ nanoparticles, for use in photocatalytic applications [87, 88], and as magnetic core–shell GNPs for biomedical applications [89, 90]. A formation model was proposed for the synthesised particles. Two components for the core and shell are present in the initial precursor droplet. The precipitation of these components depends on their individual critical saturation levels, determining the relative time of precipitation from the droplet as it evaporates and shrinks due to elevated temperatures in the furnace. The critical saturation levels depend on the physical properties (solubility, density) of the precursors for GNPs (gold chloride), for the Ti- or Fe-containing precursors, and on the concentrations and concentration ratios of these components in the droplet. In order to obtain favourable oxide core-gold shell morphologies, the initial precursor concentrations for gold need to be lower than for the oxide. In these conditions the oxide precipitates first, and forms the main volume of the composite particle. The experimental studies yielded satellite-type core–shell particles of $TiO_2@Au$ and $Fe_2O_3@Au$ with $Fe_3O_4@Au$. Additional processing and optimisation of the ultrasonic spray pyrolysis parameter conditions are required for the formation of more favourable structures with a continuous shell and other beneficial properties of these structures for catalytic and magnetic applications with a plasmonic advantage [87–91] (Fig. 3.23).

3.6.3   Hybrid GNPs with Other Noble Metals

The improved properties of hybrid GNPs are not associated with the shell of these nanoparticles, but with the synergistic effects of the structural interactions between the plasmonic and other materials in the composite. This is evident in the reports on enhanced catalytic activity for oxygen reduction processes when using these composite nanoparticles, such as with Au@Ag and Au@Pd core–shell nanoparticles [13, 121].

The SPR functionality of noble metal nanoparticles is also enhanced due to this electron synergistic effect and plasmon coupling. Au@Ag core–shell structures were synthesised with a narrow plasmon linewidth and good chemical stability, not characteristic for single Ag nanostructures [122]. Firstly, GNP bipyramids were synthesised with seed-mediated growth with cetyltrimethylammonium bromide (CTAB) as the stabiliser. An Ag shell was grown on the GNP bipyramids, with reduction of the $AgNO_3$ by ascorbic acid and cetyltrimethylammonium chloride (CTAC) as the stabiliser. The narrower SPR extinction peak and increased stability of the resulting particles are considered to be due to the changed electron density in the Ag shell

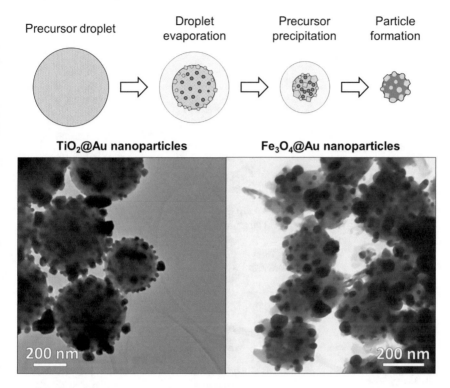

Fig. 3.23 Proposed formation model for ultrasonic spray pyrolysis synthesis of satellite-type oxide core-gold shell nanoparticles, with examples of TiO_2@Au and Fe_3O_4@Au particles produced by this method [87–90]

covering the highly electronegative Au core, or because of the charge redistribution in the Ag atom orbitals during synthesis—both a result of the electron synergistic effects [13, 122]. Electron transfer from the Au core to the Ag shell also improves the stability and silver oxidation resistivity in spherical Au@Ag particles [123].

Multicomponent noble metal core–shell particles are also an interesting subject of research, as they provide special optical, catalytic or stability properties. Au@Ag@Au core double-shell nanoparticles have greatly increased plasmon coupling while retaining the biocompatibility and stability of single GNPs. The plasmon coupling in these particles enhances the optical responses for two-photon photoluminescence, which can be used for development of two-photon excitation sensing and imaging in biomedical applications and in vivo examinations [123]. Raspberry-like Au@Pt@Au nanoparticles possess the high catalytic activity of Pt particles, while also including the SPR functionality of GNPs. The high catalytic activity is facilitated by the available relatively large surface area of the Pt shell [124].

The principles and synthesis methods of producing these nanostructures are similar to the ones described before, based mainly on the sequential reduction of

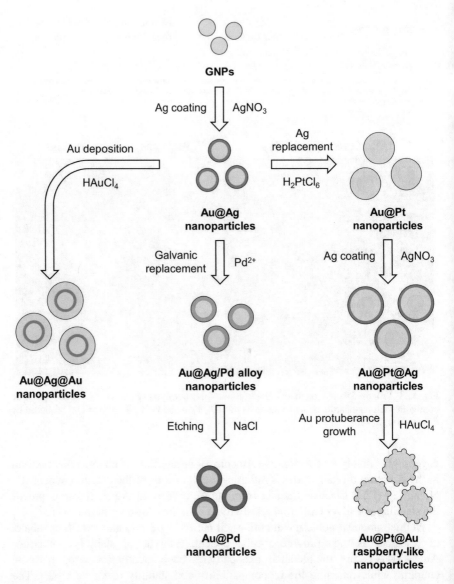

Fig. 3.24 Synthesis routes of multicomponent hybrid GNPs with noble metals: Au@Ag@Au core double-shell nanoparticles [123], core–shell Au@Pd via Au@Ag/Pd alloy structure [121], raspberry-like Au@Pt@Au nanoparticles [124]

the individual metal precursors and growth of the shell on the core surface. An additional step can be the galvanic replacement of a shell metal with another, such as replacing an Ag shell with a Pd shell [121]. Examples extend from single step to multi step synthesis, with or without cores seeds, surfactants, intermediate layers, and so on (Fig. 3.24).

3.6.4 Silica Coated GNPs

Silica presents a popular option for coating of nanoparticles, as it has high colloidal stability and dispersibility, is optically transparent, thus preserving the GNPs' SPR functionality, is chemically inert and biocompatible, often improving the cytotoxicity properties of the core it is covering. The reactive silanol groups on the surface of the silica also enable easy surface modification, and are responsible for its high stability due to electrostatic repulsion. Coating GNPs with silica improves their use in photothermal treatment, photoacoustic imaging and others [13].

Silica can also be used as an intermediate layer between a magnetic core and a plasmonic shell, as Au growth on magnetite is difficult due to their crystal lattice mismatch. The intermediate silica layer allows for GNP seeding on its surface and further growth of an Au layer [117].

GNPs with mesoporous silica shells combine the catalytic properties of the core with the shape-selective behaviour of the mesoporous shell [67]. This structure may also be modified through ordering of mesopores, controlling pore sizes, volumes and surface areas. Modifying these structures allows for loading of drugs inside the pores, used for targeted drug delivery with particles that facilitate magnetic guidance [117], or generally for enabling reactant and analyte transfer to the plasmonic core for catalysis or sensing [13].

3.7 Synthesis of GNPs through Ultrasonic Spray Pyrolysis and the Lyophilisation Process [51]

The production of GNPs on a global scale is growing rapidly, along with their increased use cases in everyday products and services in the various fields described in the previous chapters. GNP based rapid LFIA tests for the fast diagnostic detection of infection with the SARS-CoV-2 coronavirus have been used extensively since the outbreak of the virus. A daily use of several millions of these tests has, thus, also promoted the production of GNPs, as they represent the markers for control and test lines, one of the main constituents in the rapid tests, along with specific proteins for virus detection. The ability to bind GNPs with different molecules is a property that is connected to the surface state of GNPs, which is completely different from bulk gold. Namely, the surface energy has been one of the topics of atomistic research for

GNPs recently [125]. To define the thermodynamics of GNPs, it is crucial to know the surface properties and structure of these nanoparticles, which include surface energy and surface tension. In the case of spherical particles, the surface tension is considered as a force acting in the tangential plane of the nanoparticle. On the surface of a spherical particle, the coordination number, i.e. the number of atoms directly surrounding the atom, is necessarily less than inside the particle. Therefore, GNPs on the surface have different properties compared to the interior. Surface tension and surface energy are those properties that allow conjugation with various other molecules, and this option is used primarily in rapid LFIA tests.

The development for production of relevant and sensitive GNPs is essential for their widespread implementation. The different production methods offer various possibilities, as well as limitations when producing GNPs. The most favourable routes for producing GNPs with chemical reduction, such as the Turkevich or Brust-Schriffrin methods, are used extensively by research groups to produce small amounts of GNPs. These methods provide some degree of flexibility, but may be less suitable for commercial applications, such as for the manufacturing of GNPs for labels in rapid LFIA tests. For these applications, higher quantities are important, with batch-to-batch consistency regarding size, shape, uniformity, size dispersion, concentration, stability and surface modification.

Accurate data on the GNP production methods of companies that provide GNPs commercially is not easily obtained, but it can be narrowed down to two key existing methods: The chemical reduction of Au^{3+} and Au^{1+} solutions and pulsed laser ablation of Au targets. The chemical reduction is a batch process, where it is possible to produce up to 350 L in a single batch with various forms of nanoparticles, such as round, cubic, urchin or rod shaped. Pulsed laser ablation can be a batch or a continuous process, where the Au target is vaporised into a gas or liquid with a high energy laser beam, producing round or irregular shapes, as described in previous chapters [51].

Aerosol-based synthesis methods are also commercially viable for GNP production, as they are continuous and flexible processes able to produce large quantities of GNPs. Our research and development in the Ultrasonic Spray Pyrolysis method allowed for the production of monodisperse GNPs with round and distorted shapes, with high capacities for commercial use. The GNPs can be controlled easily with this method by changing the process parameters, such as the metal salt solution using Au-chloride [83], Au-acetate [84] or Au-nitrate [26], the gas flows, reaction temperatures and the collection medium with stabilisers that prevent GNP agglomeration, such as sodium citrate, polyvinylpyrrolidone (PVP) or polyethylene glycol (PEG).

The advantages of Ultrasonic Spray Pyrolysis are the continuous synthesis process, the possibility to produce homogeneous chemical compositions and multi-component particles, the relatively simple setup and operation, scalability, good nanoparticle size control and particle stability. The disadvantages are the possibility to obtain aggregated particles and not being able to use precursor solutions with high concentrations or high viscosity.

Table 3.3 Comparison of the main technological parameters of the predominant methods for commercial GNP production with Ultrasonic Spray Pyrolysis, from [51]

	Wet chemical reduction in solutions	Pulsed laser ablation	Ultrasonic Spray Pyrolysis
Size control	Good	Good	Good
Size dispersity	Low	Low	Moderate
Shape control	Spherical, rod, cubic, urchin, shell, etc.	Spherical	Spherical
Process type	Batch	Batch or continuous	Continuous
Process repeatability	High batch inconsistencies	Low	Low
Impurities	Possible unreacted contaminants	Low	Possible acidic contaminants
Scalability	Limited to max. 350 L batch sizes	Highly scalable	Highly scalable
Chemical efficiency	Moderate to High	High	Moderate
Energy intensity	Low	High	Moderate
Drawbacks	Low production rate	Possibility of cross contamination	Explosion risk (use of volatile gases)

Freeze drying, or lyophilisation, provides additional nanoparticle stability and a favourable option for controlling the GNP concentration and redispersing them in a chosen solvent, useful for various applications and commercialisation. The lyophilisation process uses freezing and drying steps to form ice crystals in the suspension, which are then sublimated from the frozen product in a vacuum, to produce a final dried product of GNPs and stabilisers. The stresses induced on the particles during freezing and dehydration may cause agglomeration of the nanoparticles and loss of product. As the product freezes, a phase separation of ice and the concentrated phase of GNPs and stabilisers occurs. This high concentration may promote irreversible aggregation of the nanoparticles. Additional substances, or cryoprotectants, protect the product from destabilisation and increase its stability during drying and storage, by immobilising the nanoparticles in a glassy matrix. The most common cryoprotectants for freeze drying are sugars, such as sucrose, glucose, or trehalose [126].

Using Ultrasonic Spray Pyrolysis with lyophilisation presents an efficient way for dry GNP production, with the preservation of the physical and chemical properties of the GNPs with their shapes, size distributions, surface modifications and long-term stability. As described previously, the GNPs produced with these methods were used in investigations for printing electronic devices [92, 97, 98], in studies for biomedical applications [85, 86, 93–95] and dental resins [96], in cosmetic products for skin care and for coatings on eyewear products. The use of Ultrasonic Spray Pyrolysis and drying with lyophilisation was demonstrated to be optimal for using the GNPs for rapid LFIA tests, as the dried GNPs can be redispersed in a suitable solvent, such as Phosphate Buffer Saline (PBS). This avoids the complex process of solvent removal or replacement during the conjugation process for the preparation of the LFIA test markers. This method was used for the development of rapid tests for detection of infection with the SARS-CoV-2 coronavirus [51].

The global production of GNPs is increasing quickly with the emergence of new products such as new Nanodevices an Nanosystems having built-in GNPs. The commercialisation and implementation of GNPs products is further promoted with the development of production methods and combination of new processes, such as Ultrasonic Spray Pyrolysis with lyophilisation for producing dried GNPs that have never been used to make nanoparticles. This represents the achievement of a solid state of GNPS, which increases the possibility of application in still unknown areas.

References

1. Mody VV, Siwale R, Singh A, Mody HR (2010) Introduction to metallic nanoparticles. J Pharm Bioallied Sci 2:282–289. https://doi.org/10.4103/0975-7406.72127
2. Pattnaik P (2005) Surface plasmon resonance. Appl Biochem Biotechnol 126:79–92. https://doi.org/10.1385/ABAB:126:2:079
3. Giljohann DA, Seferos DS, Daniel WL, Massich MD, Patel PC, Mirkin CA (2010) Gold nanoparticles for biology and medicine. Angew Chemie Int Ed 49:3280–3294. https://doi.org/10.1002/anie.200904359
4. Kodas TT, Hampden-Smith MJ (1998) Aerosol processing of materials, 1st edn. Wiley-VCH, New York
5. Khan AK, Rashid R, Murtaza G, Zahra A (2014) Gold nanoparticles: synthesis and applications in drug delivery. Trop J Pharm Res 13:1169–1177. https://doi.org/10.4314/tjpr.v13i7.23
6. Zeng S, Baillargeat D, Ho HP, Yong KT (2014) Nanomaterials enhanced surface plasmon resonance for biological and chemical sensing applications. Chem Soc Rev 43:3426–3452. https://doi.org/10.1039/C3CS60479A
7. Huang X, El-Sayed MA (2010) Gold nanoparticles: optical properties and implementations in cancer diagnosis and photothermal therapy. J Adv Res 1:13–28. https://doi.org/10.1016/j.jare.2010.02.002
8. Kelly KL, Coronado E, Zhao LL, Schatz GC (2003) The optical properties of metal nanoparticles: the influence of size, shape, and dielectric environment. J Phys Chem B 107:668–677. https://doi.org/10.1021/jp026731y
9. Mie G (1908) Beiträge zur Optik trüber Medien, speziell kolloidaler Metallösungen. Ann Phys 330:377–445. https://doi.org/10.1002/andp.19083300302
10. Bansal SA, Kumar V, Karimi J, Singh AP, Kumar S (2020) Role of gold nanoparticles in advanced biomedical applications. Nanoscale Adv 2:3764–3787. https://doi.org/10.1039/d0na00472c
11. Dalal N, Boruah BS, Neoh A, Biswas R (2019) Correlation of surface plasmon resonance wavelength (SPR) with size and concentration of noble metal nanoparticles. Ann Rev Res 5:1–6. https://doi.org/10.19080/arr.2019.05.555658
12. nanoComposix gold nanoparticles: optical properties. https://nanocomposix.com/pages/gold-nanoparticles-optical-properties. Accessed 10 Dec 2021
13. Kang H, Buchman JT, Rodriguez RS, Ring HL, He J, Bantz KC, Haynes CL (2019) Stabilization of silver and gold nanoparticles: preservation and improvement of plasmonic functionalities. Chem Rev 119:664–699. https://doi.org/10.1021/acs.chemrev.8b00341

14. Eustis S, El-Sayed MA (2006) Why gold nanoparticles are more precious than pretty gold: noble metal surface plasmon resonance and its enhancement of the radiative and nonradiative properties of nanocrystals of different shapes. Chem Soc Rev 35:209–217. https://doi.org/10.1039/b514191e

15. Zhang J, Mou L, Jiang X (2020) Surface chemistry of gold nanoparticles for health-related applications. Chem Sci 11:923–936. https://doi.org/10.1039/c9sc06497d

16. Nafisi S, Maibach HI (2017) Nanotechnology in cosmetics. In: Cosmetic science and technology: theoretical principles and applications. Elsevier, pp 337–361

17. Hu X, Zhang Y, Ding T, Liu J, Zhao H (2020) Multifunctional gold nanoparticles: a novel nanomaterial for various medical applications and biological activities. Front Bioeng Biotechnol 8:990. https://doi.org/10.3389/fbioe.2020.00990

18. Li G, Jin R (2013) Catalysis by gold nanoparticles: carbon-carbon coupling reactions. Nanotechnol Rev 2:529–545. https://doi.org/10.1515/ntrev-2013-0020

19. Carabineiro SAC (2019) Supported gold nanoparticles as catalysts for the oxidation of alcohols and alkanes. Front Chem 7:702

20. Tomić S, Đokić J, Vasilijić S, Ogrinc N, Rudolf R, Pelicon P, Vučević D, Milosavljević P, Janković S, Anžel I, Rajković J, Rupnik MS, Friedrich B, Čolić M (2014) Size-dependent effects of gold nanoparticles uptake on maturation and antitumor functions of human dendritic cells in vitro. PLoS One 9:e96584. https://doi.org/10.1371/journal.pone.0096584

21. Zhang J, Ma A, Shang L (2018) Conjugating existing clinical drugs with gold nanoparticles for better treatment of heart diseases. Front Physiol 9:642. https://doi.org/10.3389/fphys.2018.00642

22. Chen H, Kou X, Yang Z, Ni W, Wang J (2008) Shape- and size-dependent refractive index sensitivity of gold nanoparticles. Langmuir 24:5233–5237. https://doi.org/10.1021/la800305j

23. Patungwasa W, Hodak JH (2008) pH tunable morphology of the gold nanoparticles produced by citrate reduction. Mater Chem Phys 108:45–54. https://doi.org/10.1016/j.matchemphys.2007.09.001

24. Baig N, Kammakakam I, Falath W (2021) Nanomaterials: a review of synthesis methods, properties, recent progress, and challenges. Mater Adv 2:1821–1871. https://doi.org/10.1039/D0MA00807A

25. Buesser B, Pratsinis SE (2012) Design of nanomaterial synthesis by aerosol processes. Annu Rev Chem Biomol Eng 3:103–127. https://doi.org/10.1146/annurev-chembioeng-062011-080930

26. Shariq M, Friedrich B, Budic B, Hodnik N, Ruiz-Zepeda F, Majerič P, Rudolf R (2018) Successful synthesis of gold nanoparticles through ultrasonic spray pyrolysis from a Gold(III) nitrate precursor and their interaction with a high electron beam. ChemistryOpen 7:533–542. https://doi.org/10.1002/open.201800101

27. Wang W, Ding X, Xu Q, Wang J, Wang L, Lou X (2016) Zeta-potential data reliability of gold nanoparticle biomolecular conjugates and its application in sensitive quantification of surface absorbed protein. Colloids Surfaces B Biointerfaces 148:541–548. https://doi.org/10.1016/j.colsurfb.2016.09.021

28. Kim T, Lee CH, Joo SW, Lee K (2008) Kinetics of gold nanoparticle aggregation: experiments and modeling. J Colloid Interface Sci 318:238–243. https://doi.org/10.1016/j.jcis.2007.10.029

29. Ramalingam V, Revathidevi S, Shanmuganayagam TS, Muthulakshmi L, Rajaram R (2017) Gold nanoparticle induces mitochondria-mediated apoptosis and cell cycle arrest in nonsmall cell lung cancer cells. Gold Bull 50:177–189. https://doi.org/10.1007/s13404-017-0208-x

30. Moreira AF, Rodrigues CF, Reis CA, Costa EC, Correia IJ (2018) Gold-core silica shell nanoparticles application in imaging and therapy: a review. Microporous Mesoporous Mater 270:168–179. https://doi.org/10.1016/j.micromeso.2018.05.022

31. Mu L, Sprando RL (2010) Application of nanotechnology in cosmetics. Pharm Res 27:1746–1749. https://doi.org/10.1007/s11095-010-0139-1

32. Linic S, Aslam U, Boerigter C, Morabito M (2015) Photochemical transformations on plasmonic metal nanoparticles. Nat Mater 14:567–576. https://doi.org/10.1038/nmat4281

33. Narayanan R, El-Sayed MA (2005) Catalysis with transition metal nanoparticles in colloidal solution: nanoparticle shape dependence and stability. J Phys Chem B 109:12663–12676. https://doi.org/10.1021/jp051066p
34. Thompson DT (2007) Using gold nanoparticles for catalysis. Nano Today 2:40–43. https://doi.org/10.1016/S1748-0132(07)70116-0
35. Yoshida T, Murayama T, Sakaguchi N, Okumura M, Ishida T, Haruta M (2018) Carbon monoxide oxidation by polyoxometalate-supported gold nanoparticulate catalysts: activity, stability, and temperature- dependent activation properties. Angew Chemie 130:1539–1543. https://doi.org/10.1002/ange.201710424
36. Cruz B, Albrecht A, Eschlwech P, Biebl E (2019) Inkjet printing of metal nanoparticles for green UHF RFID tags. Adv Radio Sci 17:119–127. https://doi.org/10.5194/ars-17-119-2019
37. Huang D, Liao F, Molesa S, Redinger D, Subramanian V (2003) Plastic-compatible low resistance printable gold nanoparticle conductors for flexible electronics. J Electrochem Soc 150:G412. https://doi.org/10.1149/1.1582466
38. Tan HW, An J, Chua CK, Tran T (2019) Metallic nanoparticle inks for 3D printing of electronics. Adv Electron Mater 5:1800831. https://doi.org/10.1002/aelm.201800831
39. Kang JS, Ryu J, Kim HS, Hahn HT (2011) Sintering of inkjet-printed silver nanoparticles at room temperature using intense pulsed light. J Electron Mater 40:2268–2277. https://doi.org/10.1007/s11664-011-1711-0
40. Hu C, Bai X, Wang Y, Jin W, Zhang X, Hu S (2012) Inkjet printing of nanoporous gold electrode arrays on cellulose membranes for high-sensitive paper-like electrochemical oxygen sensors using ionic liquid electrolytes. Anal Chem 84:3745–3750. https://doi.org/10.1021/ac3003243
41. Deng M, Zhang X, Zhang Z, Xin Z, Song Y (2014) A gold nanoparticle ink suitable for the fabrication of electrochemical electrode by inkjet printing. J Nanosci Nanotechnol 14:5114–5119. https://doi.org/10.1166/jnn.2014.7208
42. Ali ME, Hashim U, Mustafa S, Che Man YB, Islam KN (2012) Gold nanoparticle sensor for the visual detection of pork adulteration in meatball formulation. J Nanomater 2012:1–7. https://doi.org/10.1155/2012/103607
43. Fang C, Dharmarajan R, Megharaj M, Naidu R (2017) Gold nanoparticle-based optical sensors for selected anionic contaminants. TrAC Trends Anal Chem 86:143–154. https://doi.org/10.1016/j.trac.2016.10.008
44. Nath N, Chilkoti A (2002) A colorimetric gold nanoparticle sensor to interrogate biomolecular interactions in real time on a surface. Anal Chem 74:504–509. https://doi.org/10.1021/ac015657x
45. Rastogi S, Kumari V, Sharma V, Ahmad FJ (2021) Gold nanoparticle-based sensors in food safety applications. Food Anal Methods 1:1–17. https://doi.org/10.1007/S12161-021-02131-Z/TABLES/2
46. Chang C-C, Chen C-P, Wu T-H, Yang C-H, Lin C-W, Chen C-Y (2019) Gold nanoparticle-based colorimetric strategies for chemical and biological sensing applications. Nanomaterials 9:861. https://doi.org/10.3390/nano9060861
47. James JZ, Lucas D, Koshland CP (2012) Gold nanoparticle films as sensitive and reusable elemental mercury sensors. Environ Sci Technol 46:9557–9562. https://doi.org/10.1021/es3005656
48. Stuchinskaya T, Moreno M, Cook MJ, Edwards DR, Russell DA (2011) Targeted photo-dynamic therapy of breast cancer cells using antibody–phthalocyanine–gold nanoparticle conjugates. Photochem Photobiol Sci 10:822. https://doi.org/10.1039/c1pp05014a
49. Amina SJ, Guo B (2020) A review on the synthesis and functionalization of gold nanoparticles as a drug delivery vehicle. Int J Nanomedicine 15:9823–9857. https://doi.org/10.2147/IJN.S279094
50. Aminabad NS, Farshbaf M, Akbarzadeh A (2019) Recent advances of gold nanoparticles in biomedical applications: state of the art. Cell Biochem Biophys 77:123–137. https://doi.org/10.1007/s12013-018-0863-4

51. Jelen Ž, Majerič P, Zadravec M, Anžel I, Rakuša M, Rudolf R (2021) Study of gold nanoparticles' preparation through ultrasonic spray pyrolysis and lyophilisation for possible use as markers in LFIA tests. Nanotechnol Rev 10:1978–1992. https://doi.org/10.1515/ntrev-2021-0120

52. Raj S, Sumod U, Jose S, Sabitha M (2012) Nanotechnology in cosmetics: opportunities and challenges. J Pharm Bioallied Sci 4:186. https://doi.org/10.4103/0975-7406.99016

53. Jagannathan A, Rajaramakrishna R, Rajashekara KM, Gangareddy J, Pattar KV, Rao SV, Eraiah B, Angadi VJ, Kaewkhao J, Kothan S (2020) Investigations on nonlinear optical properties of gold nanoparticles doped fluoroborate glasses for optical limiting applications. J Non Cryst Solids 538:120010.https://doi.org/10.1016/j.jnoncrysol.2020.120010

54. Kaul S, Gulati N, Verma D, Mukherjee S, Nagaich U (2018) Role of nanotechnology in cosmeceuticals: a review of recent advances. J Pharm 2018:1–19. https://doi.org/10.1155/2018/3420204

55. Oro gold cosmetics: 24 K nano gold technology from oro gold cosmetics. http://orogoldreview.blogspot.com/2013/07/24k-nano-gold-technology-from-oro-gold.html. Accessed 23 Dec 2021

56. Sengul AB, Asmatulu E (2020) Toxicity of metal and metal oxide nanoparticles: a review. Environ Chem Lett 18:1659–1683. https://doi.org/10.1007/s10311-020-01033-6

57. Wu Y, Ali MRK, Chen K, Fang N, El-Sayed MA (2019) Gold nanoparticles in biological optical imaging. Nano Today 24:120–140. https://doi.org/10.1016/j.nantod.2018.12.006

58. Piludu M, Medda L, Monduzzi M, Salis A (2018) Gold nanoparticles: a powerful tool to visualize proteins on ordered mesoporous silica and for the realization of theranostic nanobioconjugates. Int J Mol Sci 19:1991. https://doi.org/10.3390/ijms19071991

59. Nguyen DT, Kim D-J, Kim K-S (2011) Controlled synthesis and biomolecular probe application of gold nanoparticles. Micron 42:207–227. https://doi.org/10.1016/j.micron.2010.09.008

60. Lung J-K, Huang J-C, Tien D-C, Liao C-Y, Tseng K-H, Tsung T-T, Kao W-S, Tsai T-H, Jwo C-S, Lin H-M, Stobinski L (2007) Preparation of gold nanoparticles by arc discharge in water. J Alloys Compd 434–435:655–658. https://doi.org/10.1016/j.jallcom.2006.08.213

61. Mafuné F, Kohno J, Takeda Y, Kondow T, Sawabe H (2001) Formation of gold nanoparticles by laser ablation in aqueous solution of surfactant. J Phys Chem B 105:5114–5120. https://doi.org/10.1021/jp0037091

62. Menéndez-Manjón A (2009) Mobility of nanoparticles generated by femtosecond laser ablation in liquids and its application to surface patterning. J Laser Micro/Nanoeng 4:95–99. https://doi.org/10.2961/jlmn.2009.02.0004

63. Prasad Yadav T, Manohar Yadav R, Pratap Singh D (2012) Mechanical milling: a top down approach for the synthesis of nanomaterials and nanocomposites. Nanosci Nanotechnol 2:22–48. https://doi.org/10.5923/j.nn.20120203.01

64. Lee K-M, Park S-T, Lee D-J (2005) Nanogold synthesis by inert gas condensation for immuno-chemistry probes. J Alloys Compd 390:297–300. https://doi.org/10.1016/j.jallcom.2004.08.047

65. Corbierre MK, Beerens J, Lennox RB (2005) Gold nanoparticles generated by electron beam lithography of Gold(I)−thiolate thin films. Chem Mater 17:5774–5779. https://doi.org/10.1021/cm051085b

66. Birtcher RC, Donnelly SE, Schlutig S (2004) Nanoparticle ejection from gold during ion irradiation. Nucl Instrum Methods Phys Res Sect B Beam Interact with Mater Atoms 215:69–75. https://doi.org/10.1016/S0168-583X(03)01789-0

67. Zhao P, Li N, Astruc D (2013) State of the art in gold nanoparticle synthesis. Coord Chem Rev 257:638–665. https://doi.org/10.1016/j.ccr.2012.09.002

68. Sengani M, Grumezescu AM, Rajeswari VD (2017) Recent trends and methodologies in gold nanoparticle synthesis—a prospective review on drug delivery aspect. OpenNano 2:37–46. https://doi.org/10.1016/j.onano.2017.07.001

69. Alaqad K, Saleh TA (2016) Gold and silver nanoparticles: synthesis methods, characterization routes and applications towards drugs. J Environ Anal Toxicol 6:1–10. https://doi.org/10.4172/2161-0525.1000384

70. Turkevich J, Stevenson PC, Hillier J (1951) A study of the nucleation and growth processes in the synthesis of colloidal gold. Discuss Faraday Soc 11:55. https://doi.org/10.1039/df9511 100055
71. Kimling J, Maier M, Okenve B, Kotaidis V, Ballot H, Plech A (2006) Turkevich method for gold nanoparticle synthesis revisited. J Phys Chem B 110:15700–15707. https://doi.org/10.1021/jp061667w
72. Hammami I, Alabdallah NM, Al JA, Kamoun M (2021) Gold nanoparticles: synthesis properties and applications. J King Saud Univ Sci 33:101560. https://doi.org/10.1016/j.jksus.2021.101560
73. Dong J, Carpinone PL, Pyrgiotakis G, Demokritou P, Moudgil BM (2020) Synthesis of precision gold nanoparticles using Turkevich method. KONA Powder Part J 37:224–232. https://doi.org/10.14356/kona.2020011
74. Priecel P, Adekunle Salami H, Padilla RH, Zhong Z, Lopez-Sanchez JA (2016) Anisotropic gold nanoparticles: preparation and applications in catalysis. Chin J Catal 37:1619–1650. https://doi.org/10.1016/S1872-2067(16)62475-0
75. Tyas KP, Maratussolihah P, Rahmadianti S, Girsang GCS, Nandiyanto ABD (2021) Review: comparison of gold nanoparticle (AuNP) synthesis method. Arab J Chem Environ Res 08:436–454
76. Park J-E, Atobe M, Fuchigami T (2006) Synthesis of multiple shapes of gold nanoparticles with controlled sizes in aqueous solution using ultrasound. Ultrason Sonochem 13:237–241. https://doi.org/10.1016/j.ultsonch.2005.04.003
77. Gadge AN, Wadher SJ, Landge AD (2020) Gold nanoparticle. Int J Sci Healthc Res 5:21
78. Epifani M, Giannini C, Tapfer L, Vasanelli L (2004) Sol-gel synthesis and characterization of Ag and Au nanoparticles in SiO_2, TiO_2, and ZrO_2 thin films. J Am Ceram Soc 83:2385–2393. https://doi.org/10.1111/j.1151-2916.2000.tb01566.x
79. Gautam M, Kim JO, Yong CS (2021) Fabrication of aerosol-based nanoparticles and their applications in biomedical fields. J Pharm Investig 51:361–375. https://doi.org/10.1007/s40005-021-00523-1
80. Athanassiou EK, Grass RN, Stark WJ (2010) Chemical aerosol engineering as a novel tool for material science: from oxides to salt and metal nanoparticles. Aerosol Sci Technol 44:161–172. https://doi.org/10.1080/02786820903449665
81. Ullmann M, Friedlander SK, Schmidt-Ott A (2002) Nanoparticle formation by laser ablation. J Nanoparticle Res 4:499–509. https://doi.org/10.1023/A:1022840924336
82. Danks AE, Hall SR, Schnepp Z (2016) The evolution of 'sol–gel' chemistry as a technique for materials synthesis. Mater Horizons 3:91–112. https://doi.org/10.1039/C5MH00260E
83. Majerič P, Jenko D, Friedrich B, Rudolf R (2017) Formation mechanisms for gold nanoparticles in a redesigned ultrasonic spray pyrolysis. Adv Powder Technol 28:876–883. https://doi.org/10.1016/j.apt.2016.12.013
84. Shariq M, Majerič P, Friedrich B, Budic B, Jenko D, Dixit AR, Rudolf R (2017) Application of Gold(III) acetate as a new precursor for the synthesis of gold nanoparticles in PEG through ultrasonic spray pyrolysis. J Clust Sci 28:1647–1665. https://doi.org/10.1007/s10876-017-1178-0
85. Dokić J, Rudolf R, Tomić S, Stopić S, Friedrich B, Budic B, Anzel I, Colić M (2012) Immunomodulatory properties of nanoparticles obtained by ultrasonic spray pirolysis from gold scrap. J Biomed Nanotechnol 8:528–538
86. Rudolf R, Friedrich B, Stopić S, Anžel I, Tomić S, Čolić M (2012) Cytotoxicity of gold nanoparticles prepared by ultrasonic spray pyrolysis. J Biomater Appl 26:595–612. https://doi.org/10.1177/0885328210377536
87. Bogovic J, Rudolf R, Friedrich B (2016) The controlled single-step synthesis of Ag/TiO_2 and Au/TiO_2 by Ultrasonic Spray Pyrolysis (USP). JOM 68:330–335. https://doi.org/10.1007/s11837-015-1417-5
88. Alkan G, Rudolf R, Bogovic J, Jenko D, Friedrich B (2017) Structure and Formation Model of Ag/TiO2 and Au/TiO2 Nanoparticles Synthesized through Ultrasonic Spray Pyrolysis. Metals (Basel) 7:389. https://doi.org/10.3390/met7100389

89. Majerič P, Jenko D, Friedrich B, Rudolf R (2018) Formation of Bimetallic Fe/Au Submicron Particles with Ultrasonic Spray Pyrolysis. Metals (Basel) 8:278. https://doi.org/10.3390/met 8040278

90. Majerič P, Feizpour D, Friedrich B, Jelen Ž, Anžel I, Rudolf R (2019) Morphology of Composite Fe@Au Submicron Particles, Produced with Ultrasonic Spray Pyrolysis and Potential for Synthesis of Fe@Au Core-Shell Particles. Materials (Basel) 12:3326. https://doi.org/10.3390/ma12203326

91. Majerič P, Rudolf R (2020) Advances in Ultrasonic Spray Pyrolysis Processing of Noble Metal Nanoparticles—Review. Materials (Basel) 13:3485. https://doi.org/10.3390/ma13163485

92. Rudolf R, Shariq M, Veselinovic V, Adamovic T, Bobovnik R, Kargl R, Majeric P (2018) Synthesis of gold nanoparticles through ultrasonic spray pyrolysis and its application in printed electronics. Contemp Mater 9:106–112. https://doi.org/10.7251/COMEN1801106R

93. Bekić M, Tomić S, Rudolf R, Milanović M, Vučević D, Anžel I, Čolić M (2019) The Effect of Stabilisation Agents on the Immunomodulatory Properties of Gold Nanoparticles Obtained by Ultrasonic Spray Pyrolysis. Materials (Basel) 12:4121. https://doi.org/10.3390/ma1224 4121

94. Rudolf R, Majerič P, Tomić S, Shariq M, Ferčec U, Budič B, Friedrich B, Vučević D, Čolić M (2017) Morphology, Aggregation Properties, Cytocompatibility, and Anti-Inflammatory Potential of Citrate-Stabilized AuNPs Prepared by Modular Ultrasonic Spray Pyrolysis. J Nanomater 2017:1–17. https://doi.org/10.1155/2017/9365012

95. Majerič P, Jenko D, Budič B, Tomić S, Čolić M, Friedrich B, Rudolf R (2015) Formation of Non-Toxic Au Nanoparticles with Bimodal Size Distribution by a Modular Redesign of Ultrasonic Spray Pyrolysis. Nanosci Nanotechnol Lett 7:920–929. https://doi.org/10.1166/nnl.2015.2046

96. Golub D, Ivanič A, Majerič P, Tiyyagura HR, Anžel I, Rudolf R (2019) Synthesis of Colloidal Au Nanoparticles through ultrasonic spray pyrolysis and their use in the preparation of polyacrylate-AuNPs' composites. Materials (Basel) 12:3775. https://doi.org/10.3390/ma1222 3775

97. Rudolf R, Majeric P, Golub D, Tiyyagura HR (2020) Testing of novel nano gold ink for inkjet printing. Adv Prod Eng Manag 15:358–368. https://doi.org/10.14743/apem2020.3.371

98. Tiyyagura HR, Majerič P, Bračič M, Anžel I, Rudolf R (2021) Gold inks for inkjet printing on photo paper: complementary characterisation. Nanomaterials 11:599. https://doi.org/10.3390/nano11030599

99. Menon S, Rajeshkumar S, Kumar SV (2017) A review on biogenic synthesis of gold nanoparticles, characterization, and its applications. Resour Technol 3:516–527.https://doi.org/10.1016/j.reffit.2017.08.002

100. Qiao J, Qi L (2021) Recent progress in plant-gold nanoparticles fabrication methods and bio-applications. Talanta 223:121396. https://doi.org/10.1016/j.talanta.2020.121396

101. Adabi M, Naghibzadeh M, Adabi M, Zarrinfard MA, Esnaashari SS, Seifalian AM, Faridi-Majidi R, Tanimowo Aiyelabegan H, Ghanbari H (2017) Biocompatibility and nanostructured materials: applications in nanomedicine. Artif Cells Nanomed Biotechnol 45:833–842. https://doi.org/10.1080/21691401.2016.1178134

102. Ganguly P, Breen A, Pillai SC (2018) Toxicity of nanomaterials: exposure, pathways, assessment, and recent advances. ACS Biomater Sci Eng 4:2237–2275. https://doi.org/10.1021/acsbiomaterials.8b00068

103. Carnovale C, Bryant G, Shukla R, Bansal V (2019) Identifying trends in gold nanoparticle toxicity and uptake: size, shape, capping ligand, and biological corona. ACS Omega 4:242–256. https://doi.org/10.1021/acsomega.8b03227

104. Sani A, Cao C, Cui D (2021) Toxicity of gold nanoparticles (AuNPs): a review. Biochem Biophys Reports 26:100991. https://doi.org/10.1016/j.bbrep.2021.100991

105. Kus-Liśkiewicz M, Fickers P, Ben Tahar I (2021) Biocompatibility and cytotoxicity of gold nanoparticles: recent advances in methodologies and regulations. Int J Mol Sci 22:10952. https://doi.org/10.3390/ijms222010952

106. Lu W, Yao J, Zhu X, Qi Y (2021) Nanomedicines: redefining traditional medicine. Biomed Pharmacother 134:111103. https://doi.org/10.1016/j.biopha.2020.111103
107. Sindhwani S, Chan WCW (2021) Nanotechnology for modern medicine: next step towards clinical translation. J Intern Med 290:486–498. https://doi.org/10.1111/joim.13254
108. Mikhailova EO (2021) Gold nanoparticles: biosynthesis and potential of biomedical application. J Funct Biomater 12:70. https://doi.org/10.3390/jfb12040070
109. U. S. National Library of Medicine ClinicalTrials.gov. https://clinicaltrials.gov/. Accessed 8 Jan 2022
110. Elgamily HM, El-Sayed HS, Abdelnabi A (2018) The antibacterial effect of two cavity disinfectants against one of cariogenic pathogen: an in vitro comparative study. Contemp Clin Dent 9:457–462. https://doi.org/10.4103/ccd.ccd_308_18
111. Bapat RA, Chaubal TV, Dharmadhikari S, Abdulla AM, Bapat P, Alexander A, Dubey SK, Kesharwani P (2020) Recent advances of gold nanoparticles as biomaterial in dentistry. Int J Pharm 586:119596. https://doi.org/10.1016/j.ijpharm.2020.119596
112. Gad M, Fouda S, Al-Harbi F, Näpänkangas R, Raustia A (2017) PMMA denture base material enhancement: a review of fiber, filler, and nanofiller addition. Int J Nanomedicine 12:3801–3812. https://doi.org/10.2147/IJN.S130722
113. Ban Z, Barnakov YA, Li F, Golub VO, O'Connor CJ (2005) The synthesis of core–shell iron@gold nanoparticles and their characterization. J Mater Chem 15:4660. https://doi.org/10.1039/b504304b
114. Sood A, Arora V, Shah J, Kotnala RK, Jain TK (2017) Multifunctional gold coated iron oxide core-shell nanoparticles stabilized using thiolated sodium alginate for biomedical applications. Mater Sci Eng C 80:274–281. https://doi.org/10.1016/j.msec.2017.05.079
115. Mirrahimi M, Hosseini V, Kamrava SK, Attaran N, Beik J, Kooranifar S, Ghaznavi H, Shakeri-Zadeh A (2018) Selective heat generation in cancer cells using a combination of 808 nm laser irradiation and the folate-conjugated Fe_2O_3@Au nanocomplex. Artif Cells Nanomed Biotechnol 46:241–253. https://doi.org/10.1080/21691401.2017.1420072
116. Billen A, de Cattelle A, Jochum JK, Van Bael MJ, Billen J, Seo JW, Brullot W, Koeckelberghs G, Verbiest T (2019) Novel synthesis of superparamagnetic plasmonic core-shell iron oxide-gold nanoparticles. Phys B Condens Matter 560:85–90. https://doi.org/10.1016/j.physb.2019.02.009
117. Stafford S, Serrano Garcia R, Gun'ko Y (2018) Multimodal magnetic-plasmonic nanoparticles for biomedical applications. Appl Sci 8:97.https://doi.org/10.3390/app8010097
118. Nguyen T, Mammeri F, Ammar S (2018) Iron oxide and gold based magneto-plasmonic nanostructures for medical applications: a review. Nanomaterials 8:149. https://doi.org/10.3390/nano8030149
119. Caruntu D, Cushing BL, Caruntu G, O'Connor CJ (2005) Attachment of gold nanograins onto colloidal magnetite nanocrystals. Chem Mater 17:3398–3402. https://doi.org/10.1021/cm050280n
120. Kwizera EA, Chaffin E, Shen X, Chen J, Zou Q, Wu Z, Gai Z, Bhana S, O'Connor R, Wang L, Adhikari H, Mishra SR, Wang Y, Huang X (2016) Size- and shape-controlled synthesis and properties of magnetic-plasmonic core–shell nanoparticles. J Phys Chem C 120:10530–10546. https://doi.org/10.1021/acs.jpcc.6b00875
121. Chen D, Li C, Liu H, Ye F, Yang J (2015) Core-shell Au@Pd nanoparticles with enhanced catalytic activity for oxygen reduction reaction via core-shell Au@Ag/Pd constructions. Sci Rep 5:11949. https://doi.org/10.1038/srep11949
122. Zhu X, Zhuo X, Li Q, Yang Z, Wang J (2016) Gold nanobipyramid-supported silver nanostructures with narrow plasmon linewidths and improved chemical stability. Adv Funct Mater 26:341–352. https://doi.org/10.1002/adfm.201503670
123. Daneshvar e Asl S, Sadrnezhaad SK (2020) Gold@Silver@Gold core double-shell nanoparticles: synthesis and aggregation-enhanced two-photon photoluminescence evaluation. Plasmonics 15:409–416.https://doi.org/10.1007/s11468-019-01041-5
124. Xie W, Herrmann C, Kömpe K, Haase M, Schlücker S (2011) Synthesis of bifunctional Au/Pt/Au core/shell nanoraspberries for in situ SERS monitoring of platinum-catalyzed reactions. J Am Chem Soc 133:19302–19305. https://doi.org/10.1021/ja208298q

125. Holec D, Löfler L, Zickler AG, Vollath D, Fischer FD (2021) Surface stress of gold nanoparticles revisited, Int J Solids Struct 224:111044. https://doi.org/10.1016/j.ijsolstr.2021.111044
126. Abdelwahed W, Degobert G, Stainmesse S, Fessi H (2006) Freeze-drying of nanoparticles: formulation, process and storage considerations. Adv Drug Deliv Rev 58:1688–713. https://doi.org/10.1016/j.addr.2006.09.017

Printed in the United States
by Baker & Taylor Publisher Services